NAZARETH OR SOCIAL CHAOS

To those members of the clergy, across the span of geography and time, who have understood that the Social Teaching of the Church is not an optional "extra," but a sine qua non *of the restoration in Christ of both the temporal and spiritual realms.*

Nazareth *or* SOCIAL CHAOS

Fr. VINCENT McNABB, O.P.

JOSEPH KELLY | *foreword*

CICERO BRUCE | *introduction*

and featuring Fr. Hilary Carpenter's 1953 Eulogy to Fr. McNabb

IHS PRESS

Non est enim potestas nisi a Deo

Norfolk, VA

2009

Nazareth or Social Chaos.

First published in 1933 by Burns, Oates & Washbourne, Ltd. This edition has largely preserved the spelling, punctuation, and formatting of McNabb's original 1933 edition.

The Eulogy delivered by Fr. Carpenter reprinted herein was originally published in *Life of the Spirit* and was taken from Francis Edward Nugent, *A Vincent McNabb Anthology* (London: Blackfriars Publications, 1955), vii–xiii. The quotations from Fr. McNabb on pages vi and 9 come from his articles in *Blackfriars*: respectively, "Our Aim of Truth," Vol. 1, No. 1, April 1920: 6-10 at 10, and "St. Thomas in the Court Of Appeal," Vol. 12, No. 132, March 1931: 151-56 at 156.

ISBN-13: 978-1-932528-19-0
ISBN-10: 1-932528-19-9

Library of Congress Cataloging-in-Publication Data

McNabb, Vincent, 1868-1943.
Nazareth or social chaos / by Vincent McNabb ; foreword by Joseph Kelly ; introduction by Cicero Bruce.
p. cm.
Originally published: London : Burns, Oates & Washbourne [1933]
ISBN 978-1-932528-19-0
1. Economics--Religious aspects--Catholic Church. 2. Agriculture--Religious aspects--Catholic Church. 3. Christian sociology Catholic Church. I. Title.
BX1795.E27M42 2010
261.8--dc22

2007047749

Printed in the United States of America.

IHS Press is the only publisher dedicated exclusively to the Social Teachings of the Catholic Church.

Contents

PAGE

Foreword . . . vii
Joseph Kelly

Introduction . . . ix
Cicero Bruce

Eulogy to Rev. Fr. Vincent McNabb, O.P. . . . xxiii
Very Rev. Fr. Hilary Carpenter, O.P.

NAZARETH OR SOCIAL CHAOS

The Call of Nazareth . . . 1
On Rights and Property . . . 6
The Money Muddle . . . 10
Things and Tokens . . . 13
Social Soundings . . . 17
Are We Living on Capital? . . . 21
Over-Production or Under-Consumption? . . . 25
The Farmers' Food Raid . . . 27
Cogs in the Machine . . . 29
Facts for Whitehall . . . 32
Is Patriotism Dead? . . . 37
The Sins of Avarice . . . 40
Memento Mei . . . 43
Dear Mother Earth . . . 47
Towards Hope . . . 53
Nature's Call to Work and Thrift . . . 56
Absenteeism . . . 59
Mass-Production in Agriculture . . . 63
Group Home-Colonisation . . . 68
Fifteen Things a Distributist May Do . . . 72

Notes . . . 77

"We shall hope to tell the truth, then, with such a sense of time and person as to leave no men our enemies except such as are the enemies of truth. Yet we shall seek first to tell the truth rather than to study the subtle art of adjusting it to the circumstances of time and person. We shall not hold the dangerous axiom that 'truth is the best policy,' because policy is but a means to an end; and truth is an end, not a means."

—Rev. Fr. Vincent McNabb, O.P.
Blackfriars, April 1920

Foreword

I CAN'T RECALL WHEN I FIRST CAME ACROSS VINCENT McNabb's *Nazareth or Social Chaos*; it just seems to have been with me forever—a faithful guide and refuge that I keep returning to time and again. It's not a comprehensive exposition of Catholic social teaching, nor is it a work of any self-apparent complexity, yet it comes from a man who was one of the Catholic Church's most brilliant and erudite philosophers. The genius of McNabb was that he was able to reduce theological complexities to the simplest, most self-apparent of truths, invariably condensing whole worlds of argument into a single short sentence or phrase.

Written in 1933, when he was working tirelessly to make the ideal of Distributism a reality, McNabb's call to flee from the hopelessness of urban decay in search of a new life in Christ based on frugality and ruralism was way ahead of its time.

Fashioned resolutely from the principles of *Rerum Novarum*, his uncompromisingly negative critiques of urbanism and mass consumption have anticipated many of our present, post-modern dilemmas.

They say that it's the small, simple books that begin great revolutions; *Nazareth or Social Chaos* is undoubtedly one such work.

Joseph Kelly
Editor, *The Universe Catholic Weekly*
Manchester, England

Introduction

Very Reverend Vincent McNabb, O.P., was a man admired for his integrity and vision. He was "great in many ways," observed his friend G. K. Chesterton, "mentally and morally and mystically and practically." Hilaire Belloc described his character, wisdom, and judgment as things "altogether separate from the world." Like many who knew him personally, Msgr. Ronald Knox was struck by the holiness of his presence. He was, said Knox, "the only person I have ever known about whom I have felt, and said more than once, 'He gives you some idea of what a saint must be like.'" Maurice Baring, poet and Catholic convert, wrote about the reaction of an unbeliever friend who heard McNabb preaching at Cecil Chesterton's funeral: "A poet heard you preach and told me this: / While listening to your argument unwind / He seemed to leave the heavy world behind."[1]

McNabb was born on July 8, 1868, in Portaferry, County Down, Ireland.[2] His mother was a dressmaker and his father a sea captain. Both were proud of their precocious seventh son,

1. Quotations are from a talk given to the UK League of the Kingship of Christ by its treasurer, Michael Hennessy, at St George's House, Wimbledon, on Saturday, June 15, 2002. Hennessy's talk is accessible online at www.vincentmcnabb.org/contradiction.html, accessed June 18, 2009.

2. Biographical information about Fr. McNabb is available through various online sources. Other sources include a study by Ferdinand Valentine, *Father Vincent McNabb, O.P.: Portrait of a Great Dominican* (Westminster: Newman Press, 1955) and the special issue of *The Chesterton Review* 22.1–2 (1996), dedicated to McNabb.

whom they named Joseph and reared not far from the rock that covers the bones of St. Patrick. His early education began at the diocesan seminary of St. Malachy's College, Belfast. It continued in Newcastle-upon-Tyne, England, where circumstances required the McNabbs to move when Joseph was still a boy. There young McNabb was deeply touched and forever altered by the piety and spiritual discipline of the Dominicans who ran the local parish. At the age of seventeen he joined the novitiate of the English Dominicans at Woodchester in Gloucestershire, where he studied from 1885 to 1891 and took the religious name Vincent. In 1894 he received the degree of Lector in Sacred Theology from the University of Louvain, to which he was sent after his ordination.

The greater part of his life transpired in England, where in addition to fulfilling his various duties to his Dominican province McNabb served as (among other things) professor of philosophy at Hawkesyard Priory and lecturer at the University of London. He was often observed reading the Old Testament and making notes on it in Hebrew, reading the New Testament and quoting from it in Greek, reading the works of St. Thomas and writing reflections on them in Latin. The world was never too much with him. Nor did he allow himself to be altogether comfortable in it. He slept not in a bed, but on the wooden floor of his cell. He owned not a single chair, for his waking hours were spent either standing up for the faith of his fathers or on his knees praying. A born controversialist, McNabb engaged routinely in public debate. In Hyde Park or other public venues thousands heard him holding forth in loomed habit and cobbled boots against Protestants, atheists, freethinkers or agnostic luminaries such as George Bernard Shaw.

McNabb died in 1943 of a cancerous throat. When his doctors told him the end was near, he responded faithfully: "I don't see why I should make a tragedy of this . . . . It's what I have been preparing for all my life."[1] His earthly possessions

1. Matthew Holden, "Very Rev. Vincent McNabb, O.P.," *Catholic Authors.*

had been scant in number, but two of them he prized above all the rest. One was a badge of honor given him in 1919 by Albert, the king of the Belgians, for his (McNabb's) efforts during the First World War for the relief of Belgium; the other was his ring of a Master of Sacred Theology. Before his death, and in the genuine spirit of poverty that characterized the whole of his life, he humbly surrendered both of these to his superior. McNabb spent his last days at St. Dominic's Parish in London. He was buried in St. Mary's Catholic Cemetery. . . .

Few in our time have heard of Fr. Vincent McNabb, the fervent priest from Portaferry. Nor would many today relish what he had to say in his published essays on topics social and cultural. For McNabb was no apologist for the way we live now. In truth he repudiated it. He argued prophetically that, since the Industrial, French, and Scientific Revolutions, life in the West has lapsed into a stupor of economic confusion. A leading light among the English Distributists, he maintained with pen puissant and dialectically guided by the principles of Catholic social teaching that man is something more than a mere belly to fill, that economics, properly understood, pertains not to the welfare of quasi-realities, such as "the market" or "the state," but rather to stewardship or management of the home centered round the hearth on which sit the iconic reminders of who, what, and whose we are.

In penning and proffering his views, McNabb was not seeking merely to revel in the distillation of his thoughts paginated and bound for public consumption. Nor did he see his essays as "just a student's ascetic or mystic contemplations."[1] He saw them as the very "blood-spurtings forced from the mind and heart of a priest in life's fighting line, by the pressure of defense and attack."[2] Through his "blood-spurtings," McNabb was seriously endeavoring to incite Catholics of his

com, http://catholicauthors.com/mcnabb.html, accessed June 18, 2009.

1. Fr. Vincent McNabb, *The Church and the Land* (London: 1925; Norfolk, Va.: IHS Press, 2003), 29.

2. *Ibid.*

day to rise to a challenge, if not to organize a veritable crusade. The end of that crusade, the object of McNabb's challenge to the faithful, was a return to contemplative rural life.

Behind McNabb's challenge was the history of the British landlords, who, by the eighteenth century, had become mere money-minded squires bent on competing with urban manufacturers by embarking upon schemes of "agricultural improvement." Foremost among these schemes were the enclosures, the primary intention of which was to maximize the rents of the lords' lands. From the point of view of mere agricultural efficiency, perfected methods of cultivation were a boon, insofar as they increased production of essential foodstuffs. But in the words of one of McNabb's fellow contributors to *Flee to the Fields*, the agricultural policy of the landlords resulted finally in ousting "the old small yeomen in favour of the later big tenant farmers."[1]

Though the circumstances were altogether different, independent farming in the United States suffered a comparable decline in the first half of the last century. The cause was the American dream of prosperity under the influence of which the small farmer decided to grow fewer consumable crops to feed his family and more cash crops to exchange at market for legal tender. This decision marked the beginning of the end of the American subsistence farmer, who, ultimately, became a servant to the market as well as a victim of its fluctuations.[2] As the competitive farming of cash crops evolved into a risky enterprise necessarily reliant upon scientific methods and machines that had to be purchased with either token wealth or bank credit, many whose ancestors had enjoyed relative independence in the countryside flocked to urban areas where they joined the servile ranks of proletarian wage earners.

---

1. Commander Herbert Shove, D.S.O., R.N., "The Rise and Fall of Industrialism," in *Flee to the Fields: The Faith and Works of the Catholic Land Movement* (Norfolk, Va.: IHS Press, 2003 [1934]), 30.

2. Consider, for instance, the revolt of cotton farmers in England, Arkansas, on January 4, 1931. See the report from the *Daily Telegraph* quoted in *Nazareth or Social Chaos* (27).

The enclosures, which blighted the English peasantry, inspired Irish poet Oliver Goldsmith to write his famous elegy, "The Deserted Village." Goldsmith's poem, though written nearly two centuries earlier, is the poetic analogue to McNabb's protest against the urbanization of the modern world and the enclosures, so to speak, of the modern mind. McNabb was likely familiar with the poem, as he was a scholar of English poetry.[1] Regardless, the poem's sentiments harmonize with his own. Take this passage, for instance, concerning the plight of the displaced land-worker: "If to the city sped–What waits him there? / To see profusion that he must not share; / To see ten thousand baneful arts combined / To pamper luxury, and thin mankind; / To see those joys the sons of pleasure know, / Extorted from his fellow creature's woe."[2] Or this one: "Even now, the devastation is begun, / And half the business of destruction is done, / Even now, methinks, as pondering here I stand, / I see the rural Virtues leave the land."[3] Both passages chime with McNabb's recurrent complaint against the urbanization ("destruction") of the peasantry, which was almost complete by the time of his birth.

McNabb believed there were urgent reasons why he as priest could not ignore the rampant urbanization of the Anglo-American world. For one, the sheer number of Catholics involved was significant. Between the World Wars, up to ninety-five percent of Great Britain's Catholic population, which had grown in the nineteenth-century as a result of the Oxford Movement and as an outcome of immigration in the wake of the Irish famine, was urban. During the same period in the United States, approximately eighty percent of Catholics was concentrated in industrial towns, although urban and

1. See *Geoffrey Chaucer: A Study of His Genius and Ethics* (Ditchling: Pepler and Sewell, 1934) and *Francis Thompson and Other Essays* (Ditchling: Saint Dominic's Press, 1936).
2. *The Norton Anthology of English Literature*, 7th edition, 2 vols. (New York: W. W. Norton & Co., 2000), 1:2864.
3. *Ibid.*, 2866.

rural populations were roughly equal. But the most compelling reason of all was that the smothering "incubus of industrialism" threatened to quash Catholic culture and produce living conditions "so overwhelming as to make race suicide seem the only way out."[1]

Out of the industrial revolution emerged the modern city, in which urban confluence, competition for wages, unemployment, and rationalizing appeals to theories of "overpopulation" predisposed urbanites on both sides of the Atlantic to accept as "enlightened truth" the propaganda of neo-Malthusianism. Neo-Malthusian ideology posited the notion that unrestricted reproduction was the besetting vice of the poor and ignorant. McNabb, of course, was of a different mind and spoke out, either directly or indirectly, against the principal neo-Malthusians of the early twentieth century, such as American Socialist Margaret Sanger and her English friend Marie Stopes, the controversial author of Wise Parenthood (1918) and founder (in 1921) of the Society for Constructive Birth Control and of England's first "family planning" clinic in Holloway, London.[2] McNabb was convinced and concerned that Catholics were perilously inclined by the Sangers and the Stopeses of his time to embrace values antithetical to those of the Church and to accept the logic of secularity, which he devoted his life to combating with the principles of natural law and the habits of Thomistic reasoning.

In the neo-Malthusian city, McNabb observed, "there is no possibility [wages and rents being what they are] for the average working man to have an average family" and still be called "responsible" by his peers.[3] (What McNabb considered to be an average family is unclear, but we must remember that he was the tenth of eleven children born to a poor Irish couple

1. See McNabb's essay on the "Incubus of Industrialism" in *The Church and the Land*, 91–93.

2. Note, for example, McNabb's allusion to the case of *Stopes* v. *Sutherland* in *The Church and the Land*, 160.

3. McNabb, *The Church and the Land*, 141.

"who exemplified," as Professor William Fahey has said, "the loving and capable parenthood that so often marks McNabb's social criticism."[1]) To avoid the average family, McNabb averred quite correctly, the average working man is coerced by the circumstances of his urban situation, to which there seems to be no alternative, to limit his offspring either by contraception or by abstinence – either by mortal sin, that is, or by "what is for the average individual heroic virtue."[2]

Its negative effect on what he referred to as the fundamental unit of society, the family, was not the only reason McNabb eschewed the city. Another was that he believed the city's diurnal goings-on to be a parody of real life. The city's essential flaw, he argued, is its organization around the regnant financial world of currency and credit, one of the "many worlds of tokens created by the will of man" within "the great world of things created by the Will of God."[3] The financial world of the city, the money-world, is "not a real-world or thing-world," he maintained, "but a shadow-world."[4] It is an abstract and impalpable world, to be sure. But it is nonetheless a dangerous one, an unreality of the kind described in Plato's allegory of the cave, in which truths are manufactured and responses conditioned by manipulative wardens of mind and spirit. An expert in the money-world's shadows is, according to McNabb's analysis, inclined and desirous to possess men's realities if they will but play his shadow-game, if they are willing to exchange their realities for his token wealth.

The danger, according to McNabb, lies in the fact that the pursuit of tokens results finally in avarice because no one ever knows how many dollars or pounds is sufficient. The value of a dollar or a pound, for example, is relative to what it will purchase at one particular moment in the token world of finance, unlike the absolute value of, say, a hundredweight of wheat, which

1. William Edmund Fahey, "Introduction," in *The Church and the Land*, 9.

2. McNabb, *The Church and the Land*, 142.

3. *Nazareth or Social Chaos*, 15.

4. *Ibid.*

will support a man's life for six months, now and forever, in the world of real things and absolute measures. The undue desire for token-wealth, given its relative value, can never be satisfied. Moreover, participants in the token-world are motivated by uncertainty, the root of avarice in the money-world, to acquire more than is needed. In McNabb's "greater world of realties," on the other hand, this is not the case: "If food is needed, no man desires an infinite meal; if clothing, no man desires an infinite garment; if shelter, no man desires an infinite house; if possessions, no man desires an infinite field to till." In the world of real and truly useful things a person's "being and powers of doing have a bound which sets a limit to the things he needs."[1]

For McNabb the city was nothing less than a proximate occasion for sin and consequently a hindrance to a thriving and devout Catholic population in the industrialized West. The Church, he therefore concluded, must of necessity encourage and support a back-to-the-land movement in the developed nations. It could not expect "heroic virtue" of its urban members, but it could present them with an optional way of life, a rural way of life that had sustained Catholicism for nearly two-thousand years and was continuing to sustain it where a healthy peasantry survived and where villages were not yet deserted. McNabb was convinced that on their own and far from the madding crowd the faithful stood the best chance of fully realizing and truly actualizing their destiny as creatures of God. On the land, in other words, they could best harmonize with the immemorial descant of their Catholic faith.

And what was Catholicism for Fr. McNabb but the song through which believers know what is objectively good and true? He knew that without the Church's guidance in distinguishing relative or supposed goods from objective or real ones, men and women are subject to what St. Augustine of Hippo called *libido dominandi*. It was precisely this, "the lust for power," that begot the incubus of industrialism, which exploits

1. *Ibid.*

the metaphysically rootless who wander aimlessly in the metropolitan quagmires of apparent self-sovereignty. It was the lust for power that caused industrial England to organize its economy chiefly for exports – to become the world's workshop. The tragic result of that experiment, McNabb argued in tones evoking Oliver Goldsmith, was that

> [a] people capable of freedom and of a noble life in their own homesteads . . . crowded into the factory and the mine; where deceived by loud praises of their freedom and their dignity, they [forgot] that they [were] the bondsmen not only of their paymasters at home, but of their freer, happier paymasters abroad.[1]

McNabb told his readers in no uncertain terms that the "great industrial town which had naturally fascinated his eyes of youth and dimmed his vision to the land" was not "the flower and scent of social life but the scurf and putrescence of decay."[2] What did he mean? He meant, to borrow a few sentences from American poet and preeminent Southern Agrarian, Allen Tate,

> that in ages which suffer the decay of manners, religion, morals, codes, our indestructible vitality demands expression in violence and chaos; . . . that men who have lost both the higher myth of religion and the lower myth of historical dramatization have lost the forms of human action; . . . that they are no longer capable of defining a human objective, of forming a dramatic conception of human nature; . . . that they capitulate from their human role to a series of pragmatic conquests which, taken alone, are true only in some other world than that inhabited by men.[3]

He meant, in fine, that industrialism was the mother of secularism, which he, no less than Tate, considered the bane of the West.

There is no indication in his writings or in those about him that McNabb was familiar with Tate.[4] But he certainly

1. McNabb, *The Church and the Land*, 93.

2. *Ibid.*, 32.

3. Allen Tate, *Essays of Four Decades* (Chicago: Swallow Press, 1968), 544.

4. For an introduction to the life and work of Tate, see Cicero Bruce, "The

shared his view, at least in principle, that secularism gave us (in the words of Charles Baudelaire) the *fourmillante Cité*. The *fourmillante Cité* is a world of getting and spending, a nervous world in which men and women huddle anxiously together in the din of urban life, a mechanized world in which human dignity is subordinated to the means of production, and the means of production aimed at producing a quantity of things with little or no relation to the quality of life. The *fourmillante Cité* is, in truth, a society of material means without spiritual ends, and, in the present age of technology, it "so multiplies the means, in the lack of anything better to do," wrote Tate, "that it may have to scrap the machines as it makes them, until our descendants will have to dig themselves out of one rubbish heap after another and stand upon it, in order to make more standing room."[1] Then, said Tate, the natural world will be "literally as well as morally concealed from the eyes of men."[2]

The essays of both McNabb and Tate remind us that secularism caters to our vanity and that those who control the means and ends of production would have us ever dissatisfied with our material comforts and unceasingly pursuing, because we "deserve" them, superior material satisfactions that lie in some imagined future. The social engineers would have us believe that moral progress is inevitable and inextricably connected to the physiological development of the human race. But, as Tate warns us in words that echo the sentiments of McNabb, the future that the materialist promises is nothing more than a "naturalistic Utopia of mindless hygiene and Tom Swift's gadgets." And what the godless social engineer promises is "irresponsible perfection in the future, to be gained at the slight cost of our present consent to extinguish our moral natures in a group mind."[3]

McNabb wanted posterity to remember that, after only two hundred years of industrialization, the urban centers of

---

Stand of Allen Tate," in *Modern Age: A Quarterly Review* 42 (2000), 331–45.

1. Tate, *Essays of Four Decades*, 10.

2. *Ibid.*

3. *Ibid.*, 144.

England, Scotland, Wales, and the United States had been "reduced to such a state of economic bankruptcy . . . that race suicide could be made the only practical agenda for the people."[1] Until the end of his earthly sojourn he made it his apostolic mission to embolden as many Catholics as possible to flee to the fields. "Let your Exodus be after the coming out of Egypt," he told them. "Leave the garden cities and the flesh pots, not in order to scorn suburbia or to lead a simple life, but to worship God." Forsake your cosmopolitan neighbors "not because you hate them or despise them, but because you love them so much as to hate the conditions which degrade and enslave them."[2]

Were not the major movements of collective moral reform, particularly in the history of the Jews, movements out of mechanized, anxious city life to life on the land, if not in the desert, where God's creatures could commune with God in quiet calm? Did not Providence make "land-work the primary duty and need of beings demanding daily bread to keep them in their being?" Did not Christ choose his disciples "from work of secondary need such as fishing or of still less need such as tax collecting"? Was not land-work, in other words, "an institution so indispensible and divine that from it he took no workers, but only the wisdom of parables"? Is not Nazareth–that "highland hamlet, whose every stone was hallowed by thirty years of God's redemptive love," where "a family of families gathered together in aid and defense of life," within the circuit of which "dwelt the little self-sufficing group of land-workers and hand-workers"–the representative unit, the very epitome, of human society rightly conceived? Is not any society destined to culminate in "another Golgotha" if its people remain long removed from the land, where the social requirements of "property, chastity, and authority" coalesce "in their natural soil and setting"? These were the premising

1. McNabb, *The Church and the Land*, 32.

2. *Ibid.*, 33.

questions at the crux of McNabb's economic mission, questions which McNabb answered in the affirmative, for he was certain that rural life was "one of the necessities – one of those conditional necessities, to use the phrase of the Dumb Ox of Aquin – of the enterprise of Redemption."[1]

Professor Fahey has observed that, "between 1926 and 1930, 14,000 men formally applied for small-holding grants with the British Ministry of Agriculture."[2] How many of these men were directly induced to do so by the Catholic Land Movement spearheaded by McNabb is a matter of conjecture. But the number was likely considerable, given the movement's determination, methodical organization, and documented early successes. To be sure, McNabb and the English agrarians had little if anything to do with a comparable exodus which occurred in the United States from 1930 to 1932, when, according to Fahey, some 764,000 Americans took up life in the hinterlands.[3] Yet, this historical recollection reminds us that, up to the Second World War, agrarianism and decentralist thought on both sides of the Atlantic had not yet been entirely muted.

The agrarian spirit has not completely expired in our time. But it extends into our century through an unfortunate paucity of voices. The most powerful of these has been that of Kentucky poet and essayist Wendell Berry. Berry's essays take an unabashed stand for the "principles of neighborhood and subsistence."[4] Berry's, the globalists will chide, is a stand for "protectionionism," and they are right because (in Berry's words) "that is exactly what it is."[5] Berry knows, as McNabb knew, that a global economy can be realized and sustained

1. This paragraph rehearses McNabb's adumbrative remarks in the first chapter of *Nazareth or Social Chaos*, 1–5.

2. Fahey, Introduction to *The Church and the Land*, 11.

3. *Ibid.*

4. Wendell Berry, "The Total Economy," in *Wealth, Poverty, and Human Destiny*, ed. Doug Bandow and David L. Schindler (Wilmington, Del.: ISI Books, 2003), 429.

5. *Ibid.*

"only if nations and localities accept or ignore the inherent instability of a production economy based on exports and a consumer economy based on imports."[1] He reminds us in words that recall McNabb's that the so-called "free trade" is not really free at all:

> The "free trade," which from the standpoint of the corporate economy brings 'unprecedented growth,' from the standpoint of the land and its local populations, and ultimately from the standpoint of the cities, is destruction and slavery. Without prosperous local economies, the people have no power and the land no voice.[2]

It is tempting to expound the relevance of McNabb's essays to the present economic and moral situation. Here is not the place, though, to go on about America's sweatshops, now out of sight, and out of mind; about the nomadic lives of careerists in the global economy; about the allure of the workplace, which has replaced the home as the economic center of gravity; about the enthralling costs of fashionably inefficient automobiles, capacious homes, and cellular gadgetry; about the consumptive battening of corporatist greed upon the last vestiges of communal and personal autonomy; about the commodification of every aspect of man's pursuit of happiness; about the decay of tradition; about the loss of reflective leisure upon which authentic civilization depends. It is enough to say that McNabb's observations are forever relevant. Take this one as a representative point in case: "No true citizen should willingly grow richer by what makes his country grow poorer" or count his associates citizens if they "aim explicitly at their own wealth, wholly careless of their country's resultant poverty." The statement's inherent wisdom is of permanent value because here, as everywhere in his published writings, McNabb speaks from what he called "the sphere of principles, where alone discussion is possible."[3]

1. *Ibid.*, 428.
2. *Ibid.*, 430.
3. McNabb, *Nazareth or Social Chaos*, 37, 38.

Suffice it to say that the current importance of McNabb's commentary lies in the vexing questions it causes the thoughtful to ask—yet again. And the imperative question is, "where are we allowing ourselves to be led by our passion for industrial progress and for an ever-higher standard of living"? This, said Daniel Boorstin, "remains among the deepest, the most embarrassing, and the most unasked questions of our day."[1] It is, however, the question pertinent to any latter-day appreciation of McNabb's writings. In them we see the image of a greater and better world, and as the seventeenth-century Yorkshire divine Thomas Burnet wrote, "it is helpful sometimes to contemplate in the mind, as on a tablet, the image of a greater and better world, lest the intellect, habituated to the petty things of daily life, narrow itself and sink wholly into trivial thoughts."[2]

Cicero Bruce
Professor of Literature
Southern Catholic College
Dawsonville, Georgia
October 2, 2009
Feast of the Holy Guardian Angels

1. Quoted in Bruce, "The Stand of Allen Tate," 333.
2. See the note on Samuel Taylor Coleridge's epigraph to "The Rime of the Ancient Mariner" in *Norton Anthology*, 2:422–23.

Eulogy to Rev. Fr. Vincent McNabb, O.P.

Preached in London on the
tenth anniversary of his death

> "Lord, Thou knowest all things, Thou knowest that I love Thee."
>
> —St. John, xxi:17

KNOWING HIS END TO BE DRAWING NEAR – IT WAS IN fact the day before his death – Fr. Vincent McNabb was concerned to make certain provisions for his funeral. Calling one of the younger brethren to him, he said to him:

> Dear Father, I may well be dead tomorrow and there is a service I would ask of your charity. I don't want a shaped and polished coffin such as they usually provide, nor should I like to have a brass factory-made cross on it nor be labelled with a brass label. I want an ordinary box made of the same sort of wood as this floor

and he pointed with a smile to the common deal floor, uneven and knotted, which he had swept day by day with his bare hand, a floor such as had been his only bed for forty years and more.

> You will want some measurements, of course; go and get a tape and measure me now; leave a bit extra length for me to stretch and plenty of room for the shoulders.

> When I am in the box and have been taken down to the church, bring away the lid and get out your brushes and black pigment and paint a cross on it, a good big one. Then you will want the inscription; I will dictate it to you: "Ven. et Adm. R.P.F. Vincentius McNabb, O.P., S.T.M., Natus 8 Iulii 1868, Professus 28 Nov. 1886. Mortuus"–I don't know the date, probably tomorrow, but anyway, "1943 Londini." Now after that I want you to put a Greek inscription.

He then recited by heart these words:

> Κύριε, πάντα σὺ οἶδας, σὺ γινώσκεις ὅτι φιλῶ σε

They are the words of my text: "Lord, thou knowest all things, thou knowest that I love thee." "Then," he continued,

> after the *Requiem* and the *Libera*, to Kensal Green. I don't wish to be taken there in a glass-house. Borrow the builder's lorry and let down the sides. Put me on it, and let the two acolytes sit one on either side with their candles. (Don't light the candles; they will only get blown out.) Have the boy with the processional cross with his back to the driver's cab, and let me be driven to Kensal Green like that. Of course, I know what some people will say: "That's McNabb and his tomfoolery, McNabb and his publicity, showing off." But it isn't that, my dear Father, it isn't that. All my life I have preached and when I am no longer alive I shall still preach. I shall preach even with my dead body . . . . Now of course I realise that I have a vow of obedience and you will need the Prior's permission to do all this.

In these his last hours Fr. Vincent was meeting death as he had lived his whole life. In the incident which I have related to you we may truly see a summary of his whole approach to his religious and apostolic life. His humour was there and his humanity, his obedience, his zeal for the preaching of the word of God, his deep sincerity and dislike of the sham, and above all as a central theme his profound and burning desire to love and serve our divine Lord. "Do me a last favor, dear Father," he said to the same young priest. "Read to me the Passion of our dear Lord according to St. Mark."

From his early youth he was absorbed in the pursuit of truth and its understanding, for he realised that truth wherever he might find it would help him to a knowledge and deeper understanding of the ultimate and eternal truth which is God himself. As a young Dominican, he told us, he would try with all his power to use the intelligence God had given him, to think. "Think of anything," he would say to us, "but for God's sake think." He did not use that phrase lightly, as so many of us might. He meant literally that our thinking was for God's sake. All his own thought led him to God, and the fruit of his thinking was in every word he spoke or wrote.

But not even the ultimate truth, which is God himself, was for Fr. Vincent something merely to be known. Truth was to be known in order that it might be loved; it is the greater knowledge of God that can best lead to a greater love of God.

As the inevitable outcome of this intense love of God in Fr. Vincent came the scarcely less intense desire in him to bring all men, as far as in him lay, to this same knowledge and so to this same love of God. This love was a burning flame in him; it drove him into the highways and byways seeking souls; whether in the pulpit or in the public park, in the drawing-rooms of the rich or (as he preferred) in the houses of the poor, amongst university professors or working men, with Catholics or with those outside the fold, the truth he loved would be told and his very love of it seemed to show him the way. He was what is called a "popular" speaker, and he was so in every sense of the word except in the modern accepted sense. He was always concerned to speak to the people, and it was his peculiar gift that he could speak intelligibly to people in every walk of life. But it was his method, not his matter, that varied according to the needs and capacity of his hearers. No matter what the audience, it was the truth, God's truth, that he told. He did not fear to speak to children on the profound mysteries of the Faith, nor did the childlikeness that remained with this Master of Theology to the end fail to win their interest or minister to their understanding. A true son of St. Dominic, Fr. Vincent was, like him, above all *Doctor Veritatis*—a Teacher of Truth.

Few priests have made a greater or more lasting impact on this country as a whole during the present century than Fr. Vincent McNabb. He became something of a legend even in his own lifetime. He may have seemed to some to have been given overmuch to paradox, seemed something of a living paradox himself. This was partly because of his own uncompromising sincerity which stood out so forcibly in an age so much committed to what is sham and unreal. But it was still more because he was so well fitted by nature and grace to appreciate and to reflect the supreme paradox of Divine Truth Incarnate, Divine Truth which is not merely a Divine Idea but also a Divine Person, the Word made flesh, the Word which moves both in mind and heart and which, so freely received, must be no less freely given to others.

He was an outstanding product of his family background and upbringing where the deep, consuming faith of the Irish was grafted on a practically-minded and naturally philosophical stock. For him a Dominican vocation was the obvious, one might almost say the inevitable, one. The teaching of St. Thomas Aquinas inspired by the apostolic spirit of St. Dominic was wholly attractive and satisfying to his truth-loving and ardent character. The Order to which he knew himself to be called in his youth, and in which he was to prove himself so shining a light in his later years, is properly called the Order of Preachers, and its motto is the single word *Veritas*—Truth.

In him there was a rocklike and unquestioning faith and an unswerving loyalty to the teaching of the Church. Yet he knew that he must be prepared to give a reason for the faith that was in him, must be prepared to defend that loyalty. He knew too, none better, how the visible things of creation could help make manifest the invisible things of God. It was with this in his mind that he devoted himself to the deep study of philosophy, of theology and of Holy Scripture. Above all he steeped himself in the Gospel till it became alive for him and he found understanding and love of the Word of God in the Incarnation, found too his own divine commission to preach and to teach.

"Lovest thou me more than these?" asked our Lord of St. Peter for the third time. "Lord, thou knowest all things," answered Peter; "thou knowest that I love thee." And our Lord said to him: "Feed my sheep." It was this greater love that gave to St. Dominic the power and fruit that was in his preaching, moved men to call him "another Christ" for the mind and heart of Christ that was in him. It is this greater love alone that guarantees the power and the fruit in the preaching of his sons, the Friars Preachers. And it was certainly the source of that most fruitful apostolate of the great Friar Preacher whose memory we recall today. When he was quoting the Greek text, Fr. Vincent asked the young Father to verify it in the *Codex Sinaiticus*, of which there was a copy in the library.

> That early text will help to confirm my opinion that St. Peter did not say "Thou knowest *that* I love thee" but "Thou knowest *if* I love thee." I am at the end of a long life and even now I do not know if I have loved our Lord as I ought.

But, after our Lord's reply, Peter could have had no doubt, for he was given the commission to feed the flock of the Good Shepherd, a commission given only to those who have that greater love. And that same commission was given so surely and so fruitfully to Fr. Vincent.

Why are we concerned to keep his memory green? There are those who believe that Fr. Vincent was a saint. We, his brethren, who knew him so intimately, though we would not by one word seem to arrogate to ourselves what is in the sole judgment of the Church, we saw in him at any rate this intense love of our Lord and of souls. We knew the almost unbelievable austerities of his daily life, his love of his brethren; we knew his personal humility, we whose feet he had kissed when he thought he had offended us, or worse still for us, when we had done him some hurt. In this learned Master of Theology and powerful preacher there was the simplicity of a child. While he was already a dying man prayers were asked for him from the Rosary Guild. They offered their Holy Communions, said rosaries, made visits to the Blessed Sacrament, and all these were inscribed upon parchment in the form of a Spiritual Bouquet. The Father concerned with the making of this looked for Fr. Vincent and found him in the library. "I have brought you a Spiritual Bouquet from the Rosary Guild, Father," he said. Fr. Vincent came to him, with that wonderful smile, knelt at his feet and received the offering in his outstretched hands, while the tears of gratitude coursed down his lined face.

So many of us already owe him so much. We can best repay our debt by ensuring, in whatever way we can, that the good he wrought is not interred with his bones. There is only one other thing he himself would ask of us, as he so often asked in his lifetime, that we should pray for his soul. And we, who trust that the Lord he loved has taken him into the eternal happiness of heaven, nevertheless will make that prayer for him. May his great soul, and the souls of all the faithful departed, through the mercy of God rest in peace. Amen.

Very Rev. Fr. Hilary Carpenter, O.P.
Provincial, English Province of the
Order of Preachers
St. Dominic's, London
June 17, 1953

NAZARETH OR SOCIAL CHAOS

The Call of Nazareth

Some Thoughts on the Age-Old Exodus From the City to the Land

IT IS THE BEST PART OF TWENTY YEARS SINCE THE MATTER of this letter was first broached between us. Student and teacher were so akin in aim and ways of thinking that at the end of our thought-gathering, neither of us could measure, in the yield of thought, what was his share.

Early in our thinking—and Jesus, the Messiah of Jewry, was always the beginning and goal of our thought—we realized that the great Jewish movements of reformation and redemption were movements out of complex, organized city life to the simple life with God on the land, or even in the desert.

Gradually the dogged spadework of the archaeologist had proved to us that when Abraham left Haran for the desert it was not Chaldean slum-dwellers alone who formed his train. There was also something we can venture to call an intelligentsia, of whom Abraham was leader, in their going out from a decadent neo-paganism to the primaries of human life and liberty.

In our discussions on this earliest record of a group exodus we often asked ourselves the unanswered question: Whether it was not to this intelligentsia-led exodus that the earliest record should be assigned of an explicit and formulated Credo in an intelligent Creator. We could not see in the matchless Hexemeron[1] of the first two chapters of Genesis the product of unlettered nomads. But we were agreed that the religious atmosphere around Thare and his son Abraham at Ur and Haran[2] was such that the bugle-music of this first *Quicumque vult*[3] would be fit war-song for an intelligentsia shaking the town-dust of neo-paganism from its feet.

To us in our desperate venture of thinking there seemed a dramatic inevitability in this exodus from Chaldea being followed after some centuries by the exodus from Egypt. It was like the phenomenon of second conversion, which makes the soul's return to God authentic and final.

Again the two "goings-out" from Chaldea and Egypt were alike not merely in substance but in those lesser modes that seemed to betoken a law fulfilled. As the Chaldean exodus was led by an alarmed intelligentsia (as it seemed to us), so, too, the simple brick factory hands of Egypt were guided out into the desert with God by Moses, "skilled in the learning of the Egyptians." We could not refrain from seeing Moses surrounded by a group of intelligences who have given us a Social Code which the Greece of Solon,[4] Lycurgus,[5] Plato, and Aristotle failed to rival.

Again, it seemed to us that if the Israelitish reaction against the neo-paganism of Chaldea gave us the Hexemeron, the reaction against the neo-paganism of Egypt gave us the Decalogue. In each case reformation and inspiration came when the God-appointed leaders shepherded their people out of decadent city organization back to the land.

You will remember how it weighed upon our minds that the precedent of Abraham and Moses seemed to be set aside by Jesus Christ. Search as we would, we could not find a trace of his having left an Ur or a Memphis for the desert. Our inner conviction that all true reformation must be a return to the things primary of land-work and hand-work almost began to sicken if not die, when—*Deo Gratias*!—a text of St. Matthew restored our conviction from its sick bed.

"*Out of Egypt* have I called my Son . . . . Arise, take the Child and his Mother and go *into the land* of Israel" (Matt. ii, 15–20).

The day we found the meaning of these prophetic and inspired words we almost shouted for joy, as if a doom of doubt had been lifted from our shoulders. We almost chanted aloud the further phrase of the tax-collector Matthew: "And coming he dwelt in a hamlet called Nazareth, that it might be fulfilled which was said by the prophet, that *he shall be called a Nazarene*."

It was with joy on joy that, with still further study and prayer, we still further realized how punctiliously and completely this Son of Abraham had followed the lead of his earthly Sire by turning from the complexities and complication of city life to the simplicities and primaries of a life with God and with the earth as God has made it.

Some lesser riddles of the adventure of redemption, though still left unsolved, were so much a part of what had been solved as no longer to seem incapable of solution. Thus we had asked ourselves: If land-work is of such necessity–indeed, of primary necessity for redeeming the world–why did not the Redeemer choose all, or some, of the primary apostles from workers on the land?

The question thus broached had not long to wait for an answer. It was as if under the very poison-fang of the difficulty we found the poison's antidote. He who brought in the supernatural order did not rest it on the wreckage of the natural order:

Non eripit mortalia
Qui regna dat caelestia.
(He stealeth not the unnatural
Who giveth realms celestial.)

Indeed, he himself said, as if beforehand with an answer to our doubt: "No man putting his hand to the plough and looking back is fit for the Kingdom of God" (Luke ix, 62).[1]

1. I owe this text to one of the first band of young men who have left city life to "put their hand to the plough" at Chartridge.

The Word made flesh was not minded to disturb the Divine order which made land-work the primary duty and need of beings demanding daily bread to keep them in being. It was only from work of secondary need such as fishing or of still less need such as tax-collecting that Jesus chose his disciples. Land-work was an institution so indispensable and divine that from it he took no workers, but only the wisdom of parables.

You will remember the day when another of our lesser questions was answered to our joy. We were lopping the branches of a felled birch, to provide fuel for the bread oven. For the hundredth time we had asked ourselves, "Why did not the Son of God choose to be born on a farm?" Perhaps our previous talk on Russia's naive efforts after the Co-operative State gave us the clue. At last one of us said gravely: "But the Incarnate Word, as he could not disturb the land-unit, so he could not be born in the land-unit." To the inevitable question, "Why?" came the inevitable and satisfactory answer: "Because the divinely instituted land-unit is the normal family of father, mother, and several children. And the Word made flesh could not be One of several children." At once we saw that, as the only Child of the Heavenly Father must be the only Child of an earthly mother, he could be born nowhere save in a home whose craft did not demand the normal family of several children.

All this was gradually opening our eyes to the full meaning of the title officially given to the Redeemer of the World in the hour of the world's redemption. For ever this Son of God and of Mary, this Redeemer of the World, will be JESUS OF NAZARETH–indeed JESUS THE NAZARENE, as one says, Donald the Crofter!

I used to envy you who had made three years of Bible study in the university that is the Holy Land. How often have I made you bring back memories of the hamlets and by-ways hallowed by the feet of Jesus! But though the very stones of Nazareth were known to you, it was only after years of speech and thought in common that you, old pupil–and I, your teacher–saw what Nazareth was and meant, and what was meant by the title: Jesus the Nazarene.

For us both Nazareth was always a highland hamlet, whose every stone was hallowed by thirty years of God's redemptive love. Gradually our eyes began to see this highland hamlet as one of the necessities–one of those conditional necessities, to use the phrase of the Dumb Ox of Aquin[6]–of the enterprise of Redemption. *For Nazareth was the Unit of human society.* It was a family of families gathered together in aid and defence of life. Within its circuit dwelt the little self-sufficing group of land-workers and hand-workers.

The primary craft of land-work and the secondary yet necessary crafts of hand-work were there working together in the primary Co-operative Group. All the sanctities and social necessities of property, chastity, and authority were there in their natural soil and setting. If, then, the Son of God made Nazareth his earthly home and took it as his earthly title, it was because He, the Redeemer and the Beginning, who came to make all things new, realized that in a Nazareth alone could be the beginning of redemption.

For this reason you–my beloved pupil–and I, your unworthy teacher, have come to feel Nazareth calling us; indeed, crying out with a loud Calvary shout to us. The theme of that unceasing Nazareth cry is: "Come back, not to Ur or Memphis or Jerusalem, but to Nazareth, lest you prepare another Golgotha."

So shrill and unceasing is this call of the Nazarene that, in spite of ourselves, it is dulling our ears to all other cries of efficiency, prudence, experience, progress, statesmanship, as if they were but the cracking or rumbling of a world tottering to ruin. Much as our will shirks this challenge of the truth, we yet see that, of a truth, only Jesus of Nazareth is the Saviour and Hope of the world.

Much as our feet falter on the threshold of the way we yet know that Nazareth alone, where alone Jesus had a home, is the divine pattern to souls who covet to do the Redeemer's work in the Redeemer's way, amidst a strayed, lost people who do not yet know that their sorest need is Repentance and Redemption.

On Rights and Property

LET ME SET DOWN SOME PRINCIPLES AND DEFINITIONS, chiefly from St. Thomas Aquinas. They may help towards clearer thinking.

A Duty is an act necessitated, not by a physical, but by a moral obligation, i.e. not by nature but by Will. It is a moral, as distinct from a physical, necessity to act, e.g. the duty of living, of seeking the truth, of working, etc.

A Right is a moral, as distinct from a merely physical power to have the means necessary to fulfil a Duty. Hence Duty is primary and absolute; Right is secondary and relative to Duty.

As moral obligation and moral power presuppose a free will, Duties and Rights are predicable only of a Free Will.

The fundamental right of the free will is to be free, and not enslaved.

SERVITUS EST IMPEDIMENTUM BONI USUS POTESTATIS.[1]

Slavery is a hindrance of the good USE of power.

"To use is to apply a principle of action to its action."[2] Hence "To Use is properly an act of the will."[3] Slavery is the evil of a being that has the internal quality of free will, but has not the external condition of freedom. Hence the fundamental property or the foundation of property and ownership is our free will. A being that cannot call its will its own can call nothing its own.

1. *Summa Theologica*, Ia, IIae, Q. ii, Art.4, 3rd reply.
2. *Ibid.*, Q. xvi, art. 2.
3. *Ibid.*, Art. 1.

By liberty our acts are our own; by property our goods are our own. Free will is the psychological condition of property; and property is the material condition of freedom. "A man by his will possesses [owns] things."[1] To have (or own) is nothing else than to use or to be able to use; and this is only by an operation.[2] As a right is a moral power to use, Right is the fundamental form of property.

"OWNERSHIP . . . denotes the relation between a person and any right that is vested in him. That which a man owns is a right."[3]

Whoever, then, admits the existence of human rights has denied the arguments against human property.

The right to property does not mean that a man shall have a right to as much as he wishes, but that he shall have a right to as much as he needs.

If ownership or property is a right, and especially if ownership is moral power, as distinct from physical power, to use, it is as misleading to speak of *the* ownership of a thing as to speak of *the* right to a thing. There are as many limitations of ownership as there are limitations of right and of power.

Where there is no physical power there is no moral power; a blind man has not the right to have electric light shades.

Yet there may be no moral power where there is physical power. Physical power to use is not moral power to use; and moral power to use in one way may not be moral power to use in another way.

In one and the same material being there may be centred many physical and many moral powers to use and thus to own.

Of God alone, who alone has full physical and moral power over everything, can we say "God has *the* ownership of A or B."

All other beings have only *an* ownership in whatever they are said to own; thus, in *land* the nation has not *the* ownership

1. *Ibid.*, IIa, IIae, Q. lix, Art. 3, 1st reply.
2. *Ibid.*, Ia, IIae, Q. xxxii, Art. 1, 1st reply.
3. *Jurisprudence.* By Sir John Salmond. 7th Edit. London, p.277.

but only *an* ownership of the land. All England belongs to all England. The nation has the "altum dominium," the highest ownership of the land of the nation. But this kind of ownership by the nation is quite compatible with another kind of ownership by the individual.

Some examples may stimulate thought. (*a*) Lord B.— has a valuable picture-gallery; but is blind. His valet is an artist who has found no money in painting. The valet has the physical and psychological power to use the pictures. Lord B.—, who is said to own them, has not the power to use them, but has only the power to prevent other people from using them!

(*b*) Mr. E.—, the millionaire ship-owner, owns half a Scotch county; but his asthma confines him to the South of France. His tenants have the physical and psychological power to use his land both for tillage and for enjoyment. He can neither till nor enjoy. Almost his only ownership of the land is a power of selling; or is a selling-ownership.

(*c*) A certain Government holds the Communistic doctrine that no one can individually own land. But the individual land-workers have alone the physical power to till the land and largely the physical power to control the fruits of the land. Whether or not their very real control of the "earth and the fruits thereof" is called ownership by the central authority, it is a more effective ownership than the shadowy overlordship of a doctrinaire bureaucracy. The land-workers, who have no official ownership, can bring the land under cultivation; whereas the central Government's official ownership is a power of putting the land out of cultivation.

All this may lead us to generalize about the words Right and Property. Right is an old word with something like a simple definite meaning. Property (like Capital, Industrial, Employee, Income, etc.) is a new word with a complex or indefinite meaning.

Again, the things which are the objects of Rights and Property are of many kinds. At the one extreme there are things like food and clothing, whose use is their consumption. At the other extreme there is the earth or land which can be used in a thousand ways and even abused, but cannot be consumed. Between these two extremes there are different classes of things with different ways of being used. To state man's relation to these things, with their complex and multifold ways of use, in terms of one word "Property," is to court ambiguity. The older word Right is at once more illuminative and scientific. Thus we can say, Right of Production, Right of Use, Right of Usufruct, Right of Occupation, Right of Selling, etc., etc.

"...the more the politicians, jurists and economists study the wisdom of [St. Thomas Aquinas] the more will their minds open to the lure of scientific truth."

—Rev. Fr. Vincent McNabb, O.P.
Blackfriars, March 1931

The Money Muddle

DURING A MEETING AT THE OLYMPIA, GLASGOW, MY lecture on "Unemployment and the Land" led a man in the hall to ask me if I could give a definition of what I think he called National Credit. I ingenuously confessed that I could not give a definition of National Credit. Indeed, as there seemed some hesitation in taking my word, I again confessed that I did not know what National Credit was; and indeed I did not care.

To tell the truth, I had a suspicion that National Credit had to do with currency or token-wealth. Now, I have long felt that as the way into our social quagmire was by putting second things first, and therefore first things second, our only way out of the quagmire was by putting first things first. But as currency is not a first thing or even a second thing, but only the token of a thing, to be deeply concerned with money and the money view of the world is to sink still deeper into the quagmire. Hence my indifference to all schemes based on a money-unit.

A walk through the slums, and a study of the official reports, of Glasgow had made me perhaps unduly sensitive to the futility of the money-standard of civilization. I do not mean that I was depressed because Glasgow was Glasgow and not, say, Birmingham or Liverpool; but because, like Birmingham and Liverpool, Glasgow has a large number of human beings living in inhuman conditions through the money-muddled thinking of a small group of not ill-minded human beings,

I cannot remember all the rungs in our exchange of question and answer. I can recall that my courteous questioner said: "But there must be money paid by the man who buys the land from the State."

We were thinking of the Land Purchase[7] scheme which transformed Ireland from a country of a few large landholders to a country of many peasant-proprietors. To his question I answered a little sharply: "No. The proprietor who buys or borrows from the State need not pay in currency, but in kind." I remembered that a most efficient sliding-scale of payment was to be found in the tithe, and in the metayer system; whereby the land-worker gave a fixed proportion of his yearly harvest. This direct dealing of the borrower and lender, with its direct exchange, was more economic and efficient than another which demanded a third person who gave to the land-worker, in exchange for his tithe, a currency token which the land-worker gave to the State.

The truth was that money, which was invented to facilitate barter and measure price, has ended by darkening counsel and measuring value. Money, being a token and not a thing, could be a means and a measure, but could never be the means and the measure. Here a principle of St. Thomas was golden: "Of things related to an end the measure is the end." Thus the measure of a poker is not its power to add lustre to a brass-finished fire-grate, but to poke and rake a fire.

Of course, as things do not measure other things, but mind measures one thing by another, a mind can employ not the true measure but a false—not an essential measure but an accidental—not the measure (which is the end) but a measure (which is not the end).

If, then, we measure a thing or commodity by something which is not its end, we do not give its absolute value but only its relative value. Yet how many modern minds, when asked the value of a commodity, think in terms of the end of the commodity? If we are asked "What is the value of a hundredweight of wheat?" we naturally say "Eight, ten, or twelve shillings." We do not say "A hundredweight of wheat will support a man's life for six months." In other words, by expressing a hundredweight in terms of currency we gave, not its value, but only its money-value. No wonder that minds accustomed to the atmosphere of currency find themselves in an intellectual money-muddle.

But the money-muddle is more than intellectual: it tends to become real. Money is a token, or artifice devised by intelligence to express a reality. But intelligence, as Lewis Carroll[8] reminds us, "can make words [or tokens] mean anything." Thus the same intelligence which today has made a piece of paper marked TEN SHILLINGS mean a hundredweight of wheat may tomorrow make it mean one pound of wheat. [N.B.—I do not understand how this is done. I and my readers who have followed the recent startling fluctuations in currency values throughout the world only know that it is done]. Hence the real value of a TEN SHILLING note may veer from a hundredweight to a pound of wheat. But the real value of a hundredweight or of a pound of wheat never veers.

How great a plight must sooner or later befall a people that has lost the art of giving things their real value, and has entrusted the commonweal to the muddled judgement of men who are experts only in money-value.

Domine! Salva nos; perimus.[9]

Things and Tokens

> And Jesus took the loaves: and when he had given thanks, he distributed to them that were sat down. In like manner also of the fishes, *as much as they would.*
>
> —St. John vi:2

Even if a man sees, as I see, that the Bible, and especially the Gospels, are the world's best handbook of Economics, he will have surprises almost every time he opens the book.

It is now only a few days since the reading of this phrase in St. John's Gospel seemed to throw more light on the present state of the world than could be thrown by a library of books on Economics.

As such an instantaneous illumination seems, like a lightning flash, to make a jagged way across the darkness it dispels, I would ask my readers to be patient with me if I cannot make what I have to say as round and lucent as the sun.

Exegetes will notice that whereas this miracle of providing food in the desert is told by Matthew, Mark, Luke, and John, this mystic phrase *as much as they would* (ὅσον ἤθεον) is given only by John. Matthew, Mark, and Luke agree in the word *they were filled* (ἐχορτάσθησαν); but they lack all mention of man's willing.

We have here an instance of God allowing the human will to condition, as it were, the divine will. This is, of course, a deep mystical fact of the soul's relationship to God. Sometimes God's power seems hindered by men's evil will, as when Jesus "could not do miracles there because of their unbelief." Here, on the contrary, the miracle of multiplication is brought to the very frontiers of man's will. This phrase of St. John is all the more striking because he more than the other three gospellers is insistent upon the Will of God. Who would expect this phrase *as much as they willed* to be found in a gospel which opened with the phrase . . . "who are born not of blood, nor of the will of the flesh, nor of the will of man, but of God?"

A sounding-line to the depths of our Blessed Lord's deference to the human will may be found in the simple ethical principle: *We cannot expect from the average person more than the average goodness.*

This principle does not mean that the average person will never rise at need to that extraordinary height of human endeavour called "heroic virtue." But while the average person will sometimes rise to heroic virtue under extraordinary circumstances, it will be found that under ordinary or average circumstances the average person will have the average goodness or badness.

This simple, yet necessary, principle underlies the Church's ascetical tradition towards what she calls "occasions of sin." The Church's wise teaching and practice rest on the fact that an "occasion of sin" is a set of circumstances wherein the average person would commit sin; or could avoid sin only by an act of heroic virtue, i.e. by more than average virtue.

Let us now follow up this undeniable general principle with an equally undeniable general fact: *All that is in the world is either* THINGS *or* TOKENS OF THINGS.

A mind or will behaves in one way when in an environment of tokens; and behaves in another way in an environment of things.

So fundamental has this fact appeared to be that scholarship is now judged by loyalty to the principle, "Verify all references." No token or sign, or evidence of a thing can be accepted as scholarship when it is possible to have the self-evidence of the thing. Hence the false showy scholarship begotten of Encyclopaedias met Bibliographies. Hence, too, the token knowledge of books to be found in experienced librarians and book-sellers. In contrast with this showy knowledge of tokens there stands the solidly-based knowledge of things. The modern growth in the physical sciences is largely a result of observing things rather than tokens of things. All laboratory thoroughness rests on the principle that direct knowledge of things is primary and knowledge of tokens is secondary.

Inside the great world of things created by the Will of God are many worlds of tokens created by the will of man. One of these is the great financial world of currency and credit, which we may call the money world. It is, of course, not a real world or thing world, but a shadow-world. Yet it is so dangerous a world that an expert in these shadows, if men will play this shadow-game, will, in the end, possess his opponents' realities.

Aristotle and Aquinas, that is the pre-Christian and Christian world, are agreed that avarice or the undue desire of these tokens tends to a certain infinity. How could it be otherwise? Even ten thousand shafts of sunlight are not the sun. The token-nothings can excite an unsatisfied indefinite desire; which can at least fill the time if not the heart of man.

Far otherwise is it in the greater world of realities. The very fullness of their reality limits man's desire. No man, unless at enmity with his reason, desires infinite things. If food is needed, no man desires an infinite meal; if clothing, no man desires an infinite garment; if shelter, no man desires an infinite house; if possessions, no man desires an infinite field to till. Man's being and powers of doing have a bound which sets a limit to the things he needs.

* * * * *

We may now venture to formulate the principle acted upon by the Eternal Wisdom when he gave the hungry his miracle-bread; and conditioned his Will to give by their will to receive. The principle may be formulated thus: *In a system mainly of things the average person may be trusted to limit his wants by his needs. But in a system mainly of tokens, the average person cannot be trusted to limit his wants by his needs.*

If the Eternal Wisdom, instead of miraculously providing bread and fishes, had provided money, St. John would have been unable to say that as much as each one wanted Jesus gave. Indeed, the one recorded complaint about mal-distribution came from the thief-Apostle, who had care of the money-tokens. The poor thief who was preparing to sell God for silver and to betray love with a kiss could hardly value this precious spikenard as a token of love. He could value it only by his tokens; and counted as worth only three hundred pence an offering which has made Magdalene the undischarged creditor of Mankind.

The present world-wide "economic blizzard" (to use the phrase of two Chancellors of the Exchequer, the late Worthington Evans[10] and Lord Snowden[11]) points ominously to a breach of our formulated principle. If our present economic system throughout the civilized world is mainly a token-system, an inevitable result will follow. The average person, in other words the majority of persons, will not be content to measure their wants by their needs. Everywhere there will be the very definite desire to have more and more token-wealth. The very uncertainty of the future value of this token will heighten and foster the desire.

Yet St. Thomas has formulated another principle, which can be ignored only by inviting the curse of war. *Pax tollitur ex hoc quod cives singuli quae sua sunt quaerunt* (Peace ceases when the citizens seek each man his own).[1] This principle becomes dramatically practical in a token-system where the majority of the citizens may be expected to seek even more than their own in their almost irresistible desire not to measure their wants by their needs.

Thus it will be found that Bethlehem and Nazareth poverty is not a defect to be remedied, but a fundamental condition of all ultimate remedy and redemption.

---

1. *Summa Theologica*, IIa - IIae, Q. clxxxiii, Art. 2, ad 3. Eng. tr.

Social Soundings

When a ship-master finds his ship befogged in unknown waters he throws the lead again and again, in order to take soundings. With something like the ship-master's zeal for accurate information of where we are and what we are coming to, I venture to set down some soundings on the social fairway.

* * * * *

M.— N.— has been appointed indexer-in-chief of a new Encyclopaedia published in the United States. Under her is a staff of women, graduates of American Universities. One of the most important functions of these women graduate indexers is to classify the contents of the Encyclopaedia under general headings ; thus, "wheat" under Agriculture. One of the most important and unremitting duties of the indexer-in-chief is to correct the classifications of her assistants. One day the chief gave an assistant a group of words, including MARRIAGE, to be classified. Later on in the day, when the group of words was returned to her, she found that her assistant had classified MARRIAGE under "SPORTS AND PASTIMES"!

This true incident was told to me some eighteen months ago. The result on my mind has been so great that I have not yet made up my mind what the incident means. But it means something horrible.

Nearly everyone laughs when they first hear this horrible incident. Then the laughter dies away into something like a gulp in the throat or a chill about the heart. This seems the adequate and normal effect.

* * * * *

As a student of St. Thomas Aquinas, I had not failed to notice his genius, not only for refusing to create new words, but also for using the old words in their old traditional meaning. I had long been perplexed by the quick changes that modern social activities had been effecting on old words, such as reason, food, faith, justice, charity, etc. The word MARRIAGE especially perplexed me.

Finding myself lately with two of the leading juniors in the Divorce Court, I asked point-blank: "Could you give me the present definition of MARRIAGE in English Law?"

I was not prepared for their hesitation in answering such a simple, relevant, official question. Had I put the question to the normal married father or mother, their hesitation would not have meant that they did not understand the essence of marriage, but that scientific definitions of their status and calling were not their duty. On the other hand, an expert is supposed to know the science of his business, and science is summarized in definitions.

The younger and cleverer of the two said enquiringly to his fellow-junior: "I suppose we cannot call it a Contract?" I said "No! But it is entered into by a contract."

He went on with a hesitation which printed words will hardly convey to the reader. At each stage of his definition-building there were side-glances towards his junior, who nodded approval.

"Marriage is–an–Association of man and woman–FOR LIFE –."

Long pause! My mind was tip-toeing with expectation of how a Divorce Court junior would finish this definition, which, up to this point, was the traditional Catholic definition of Marriage. But my expectation never forecast a finish so dramatic. He went on:

"Marriage is an Association of man and woman–FOR LIFE–terminable–only by a Court–for–adultery"!!!

I feel I am making a heavy call upon the trust of my readers by asking them to believe that I have given a word for word report of this brief dialogue with the Divorce junior; who, after a few more words, begged to be excused, as the Court awaited him. Yet my mind was at once roused to thought by these few words of definition, which would hardly have been formulated in England before the nineteenth, or even the twentieth, century, and were as characteristic of England in the twentieth century as Salisbury Minster was characteristic of England in the thirteenth century. I found a class of men, professionally dedicated to the state-craft of justice, who were more equipped for defining how the fundamental social state of marriage could be destroyed than how it could be created. It was akin to that other gruesome fact that in an age, outdoing all previous ages for its shortage of houses, there had arisen an unheard-of craft of HOUSE BREAKERS. So, too, in this age of race-suicide there had arisen within the sanctities of the Law a craft of MARRIAGE BREAKERS.

Imagine the degradation that had befallen the noble profession of Boethius[12] and More[13] when even Logic put to shame. Imagine a student of logic answering his master's demand for the definition of a house by saying: "A house is a building for men and women–which can be destroyed by an order of the Ministry of Health."

If my readers would not think I wish to raise a laugh, when I only want to stifle a sarcasm or a sob, I should liken it to the answer given by a child to the question: "What is memory?" "Please, Miss, it's what you forgets with."

* * * * *

In this age of quick changes and mass production even in sensational happenings, I am afraid my next Social Sounding must appear belated.

Said Judge Hill in the Divorce Court, October 31st, 1927: "someday, perhaps, the Law will be altered so that the Court will be given discretion to ORDER DIVORCE, even where such relief is not asked for."

He added: "The wife has apparently no real religious scruples against divorce: does not ask for it; and the consequence will be that the husband and the woman will continue to live together, and there will be more illegitimate children born."

This social sounding is so ominous that action should, if possible, be taken forthwith. If at the cast of the lead such a sounding was reported to a shipmaster, the immediate command would be "Full steam astern!" Yet the only answer to the judge's words seems to be "Full steam ahead!"

The Judge's call to "free" the culprit is all the more ominous because in the Ages of Law and Reason, which are commonly called the Ages of Faith, a Judge was looked upon as *Justitia Animata*, the incarnation of Justice. But no part of justice was so hallowed by the sanctities of law as that of contractual justice, which was based on the free, deliberate purpose of the contracting parties. If a contract was broken by one of the contracting parties it could be legally dissolved in favour of the other, innocent, unoffending party. But nowadays to such an extent has the atmosphere created by divorce legislation befogged even the supreme legal experts, the Judges, that they are now demanding legal power to dissolve contracts in favour of the offending parties. If ever such power be granted we may quite consistently expect Judges to suggest that in pity for the burglar who threatens murder the threatened shall hand over their goods, or even commit suicide!

On such a principle of contract no contract can be a valid or socially valuable instrument between man and man. Modern divorce legislation and practice will kill contractual justice. And as contractual justice is the very life of all ultimate free social institutions, divorce legislation, if it is not repealed, will ultimately kill social institutions.

Are We Living on Capital?

I do not like the word Capital. I do not think that an argument based on this word, or an arrangement based on this idea, is of much worth. Indeed, if we are to find a way out of our present economic "No Road," we must clear our Economics of many such words. Yet, as what I shall say is meant for men who still think in these ways, we use the word in a last effort to oust or maim the thing.

We, therefore, frankly profess that in our opinion, which our readers can take or leave for what it is worth, there is one main generalization about the present state of things. We say advisedly "state of things," because we are concerned mainly about the present dearth, and oncoming greater dearth of things, as distinct from tokens of things of which there is no dearth, but a glut. The main generalization about this state is that *we are living on Capital.*

Even the present activity which we see here or there, especially in England, Germany, United States, is not a problem of Production out-running Consumption; but of Consumption out-running Production. In a hundred spheres of necessary human toil we are consuming not merely more than we produce, but even more than we and our ancestors produced.

One who saw this truth as an unassailable intuition was recently asked: "You have recently been to the United States. What do you think of them?" The questioner was bewildered to hear the quiet emphatic answer: "The United States are about as drunk as we were fifty years ago." Drunkenness, as such, is consumption out-running production. For the moment it is a self-unconscious living on Capital. And as England had a spasm of prosperity after the Napoleonic wars,[14] so has the United States a spasm of prosperity by using Capital as Income–the almost excusable sin of the Prodigal Son.

This reckless living on Capital, this consumption of what should last, is flagrant in many spheres. Take our forests. The modern world has destroyed more timber in fifty years than has probably been destroyed since Hiram[15] cut down cedars for the Temple of Jerusalem, or Tutankamen[16] ordered his royal coffin. Indeed, when we realize that to produce a daily newspaper with its million or two million copies we need tons of wood-pulp, every day's issue means the consumption of a glade. Experts whose word seems of weight reckon that at our present rate of consumption the world now holds only twenty-five years' of wood. We are living on Capital.

The same spendthrift selfishness is dealing with the world's store of oil. The present writer was assured by an experienced oil-well organizer that the misuse of the world's oil supply was reckless. For the moment no one can reckon how many years oil fuel may last; but these years cannot be many at the most. Again, we are recklessly living on Capital.

The present state of the British coalfields may perhaps bring home to men's minds how coal, an exhaustible form of wealth, has been treated almost as inexhaustible capital. A little over a century of our near-sighted coal-producing has beggared our plough-lands and given us an explosive miners' problem. When our capital wealth of coal was practically treated as income by the Tyne, the Tees, the Clyde, the Taff, England had something like a drunken bout of prosperity–the prodigal's prosperity which sunsets in beggary.

And what even the less thoughtful men are seeing in our depleted coal mines they may go on to see in the rapidly depleting mines of other metals–iron, tin.

It is for this reason that in the present seeming prosperity of the United States the wiser men see the prodigal having a banquet-time by treating Capital as Income.[1] History has never seen a country receiving from other countries as much capital, in men and means, as the United States received by immigration. If every able-bodied man represents at the lowest two hundred pounds of capital, the immigration of a million men meant £200,000,000 given as capital to the United States. When to this sum in addition we add the wealth represented by the destroyed capital of the exhausted forests, oil-wells, silver and gold mines of the United States, the present prosperity need not be taken at its face value.

Even the soil of America has been so treated by rapid, exhaustive cropping and deep-ploughing that its present exhaustion is like the after-effect of exhausted Capital. With an area of uncultivated (or decultivated) land and a wealth of agricultural machines without rival in the world, and now with the world's largest market within its own frontiers, American agriculture ought to be the most prosperous in the world. Yet it is in such a state of exhaustion that for the moment it is the politician's chief home anxiety.

But we would suggest to some of our younger studious readers an even subtler form of this modern Rake's Progress.[17] Machine industrialism seemed no sooner born than it gave gifts to men. It cheapened many articles of human need or use. The cotton mills, the hosiery factories, the shoe factories have an unimpeachable record of the drop in prices that came with their coming. But whilst we are not concerned, nor disinclined, to deny this cheapening we are more inclined to ask whether this cheapening was not bought by a subtle living on Capital. When machine industrialism came in it found a social arrangement whereby *labour* could be had cheaply. In other words, labour did not demand much currency. Hence there were cheap manual work and cheap houses. In a hidden, real sense there was cheap Capital. And England especially

1. This was written in America's post-war fat years. The writer's forecast came true sooner than he thought likely.

lived on this Capital in such a subtle way as to make certain commodities cheaper, from a currency point of view. But that cheapness is now a thing of the past; nor is it likely to return.

Something akin to this has happened in the religious world. The moral standards bequeathed to us by Jewry and Christendom were a precious form of moral Capital. But today it would be fair to ask whether we have not run through our moral Capital to the point of moral exhaustion. In our passion for things new, the old Decalogue has been given the mercies of the waste heap.

The present writer sees something akin to this prodigal prosperity in the newly-created mechanical art production. The gramophone has lived murderously and suicidally on the masterpieces bequeathed by past centuries. The film has sated children in their teens with a thousand masterpieces of fictional and dramatic literature. In music and in literature, as in painting and sculpture, there is little left to purloin and squander. We have lived on Capital. We have consumed; and we are producing little or nothing for our children to consume. Ichabod.[18]

Over-Production or Under-Consumption?

> The first of these reports traces the wheat crisis to a decrease on consumption rather than to an increase in production which has only kept pace with the growth of population.
>
> —*The Times*, March 27th

THIS REPORT WAS PRESENTED TO AN INTERNATIONAL conference held in Rome, and preparatory to the second wheat conference to be held in 1932. The drawing up of the report had been entrusted to the International Institute of Agriculture.

Some reflections seem obvious.

1. It is, at least, not certain that the present wheat crisis is due to over-production. Yet the other day, when I was discussing the matter with a very prominent politician, he assumed as undeniable that modern machinery had produced too much wheat. In economic language we were suffering from over-production. This politician by assuming the over-production of wheat was but voicing a widespread and perhaps dominant opinion.

Yet this report, which though not infallible is perhaps the most authoritative statement to hand, counters other less authoritative statements by saying that, far from having over-produced beyond normal consumption, we have only kept abreast of normal consumption!

2. Even if for the sake of argument we granted that there was an over-production of wheat it would not be an economic production; and for the following reasons:

Production is for consumption; and not consumption for production.

Production is not for production's sake (as Art is not for Art's sake); nor is production for gain's sake. But production is for the sake of consumption.

But production under such circumstances of time or place that it is not economically available for consumption, is not, economically speaking, production. The factors causing mere production to be uneconomic production are two:— (1) *physical*, as distance, difficulty of transport, etc.; (2) *psychological*, as desire of gain, overhead charges, tariffs, etc.

Strangely enough, it would seem, then, when physical difficulties are lessened by modern mechanical appliances, psychological difficulties are increased.

From all this we may conclude that the more the area of consumption tends to be co-terminus with the area of production (i.e., production for consumption), the more stable will be the resulting economic state.

The Farmers' Food Raid

> Five hundred FARMERS of England, Arkansas, many of them armed, stormed the business section of the town last evening, *demanding food for themselves and their families*, and threatened to raid upon local shops unless it were provided from some source without cost. Arkansas is dominantly agricultural with cotton as its leading product. There are no Communists worth speaking about in Arkansas, but the representatives of the farmers declared today that there would be another *uprising* unless either work was provided or *Food Supplies were assured.*
>
> —*Daily Telegraph*, January 5th, 1931

IF THIS WERE NOT DEEP TRAGEDY IT WOULD BE SHEER comedy and even burlesque. Nothing in Alice in Wonderland is more grotesque than this picture of land-workers flocking into a town for–food. It is as if fisherfolk from Berwick tramped to Birmingham demanding salt water.

But these so-called "farmers" of Arkansas were not really farmers. They were really cotton growers. Now even a man from Arkansas cannot eat cotton; although no doubt Arkansas could produce all the food needed by the men, women, and children on its soil.

The best reserve these starving Arkansas farmers could have made would have been, not a reserve of money (i.e., counters) in a bank, but a reserve of Arkansas-grown food (i.e.

things) in their barns. Poor folk they seem to have neither a reserve in bank nor barn. No doubt like so many of their ilk, they parted with things from the lure of counters.

This food raid of the five hundred Arkansas farmers is another proof of a principle which, to the modern financial leaders, would seem but the hooting of owls. The principle is that mass production which usually begins by increasing, at least, the token-wealth of the mass producers, inevitably and sometimes speedily ends by decreasing to starvation-point both the real-wealth and the token-wealth of the mass producers.

Farmers arming themselves with shotguns or revolvers in order to get food out of town grocery stores, are almost an ungentle manly "I told you so" for the defeat of the mass-production economists. But we are in such bewildered years that the picture of farmers arming themselves with spade and hoe to get food out of the soil seems almost a cruel satire. These Arkansas farmers may congratulate themselves that they have produced, if not much food for eating, at least, much food for thought.

Cogs in the Machine

THE PHRASE WHICH SERVES AS TITLE TO THIS ARTICLE IS to be found in the "Interim Report of the Committee on Adult Education (Ministry of Reconstruction) on the Industrial and Social Conditions in Relation to Adult Education." The signatories of the Report in their kindliness have used it to describe the condition of the modern workman, and we have used it to describe the conditions of the signatories of the Report.

The Committee on Adult Education owes its existence and powers to the Ministry of Reconstruction. But in the Report of the Committee it is hard to see any attempt at reconstruction, except in those matters for which not reconstruction but destruction is the only fit policy.

Thus these kindly committee men, under the chairmanship of the Master of Balliol, have had the eyes to see, and almost the courage to condemn, modern industrialism. They say:

> The degradation of human beings to the position of mere *hands*, and the treatment of labour as a commodity to be bought and sold, has created a revolt in the minds of a large section of the community. The conditions of industrial life have too often outraged human personality. Long hours of labour, it is true, are economically unjustifiable; but the real objection is that they are both the cause and the consequence of conditions adverse to personal freedom.

This is Bible-truth, for which the kind-hearted Committee have our thanks. But their kind-heartedness is a little less denunciatory in the following phrases:

> We have made no specific investigation into the effects upon character and mentality of the sacrifice of sound workmanship to rapidity of production, or of work which is manifestly dishonest. Nevertheless it seems to us to be undeniable that these conditions cannot exist without reacting upon the minds and characters of those engaged in such work.

The emphatic and tragic word of this sentence is "it seems to us."

There is a noble paragraph beginning with the words, "Before the rise of the present industrial system the craftsman had pleasure in his handiwork. It was indeed the main source of his education," and ending:

> It is important that the community should realize that the specialized, mechanical, and monotonous labour which forms so great a part of our industrial activities has robbed society of one of its most powerful instruments of education . . . . Sound and careful handicraft has been superseded in a large degree by the wholesale production of things which have degraded public taste; on the other hand, the public demand continues to support the production of articles which add little or nothing to the beauty of the environment, and which sometimes do not even satisfactorily fulfil their purpose.

This is a sufficiently downright statement to stimulate a Ministry of Reconstruction into some effort towards reconstructing. But its force is deadened by an earlier statement which, try as we will, we can reconcile neither with the later statement nor with facts. The earlier statement is:

> The modern industrial system has undoubtedly solved the problem of supplying the material needs of the world's population.

We are at a loss to know how these needs can be met by the production of articles which, in the minds of the signatories of the Report, "add little or nothing to the beauty of the environment, and which sometimes do not even satisfactorily fulfil their purpose," i.e., by the wholesale production of articles which are neither beautiful nor useful.

The Report is almost up to the level of its commission (if not of its mission) in the following phrase:

> The ideal industrial system would in a large measure obliterate the sharp distinction now made between *technical* and *humane* education, for it would offer means of self-expression and development . . . and would recognize the educational value of manual processes and the influence of soundly manufactured goods upon public taste and social values.

All this is good; but not towards good. The Report suggests no change in those very industrial conditions wherein even the workers' self-expression would inevitably be of greatest value to the balance-sheet. All that the Report can, or dare, suggest is a series of eleven recommendations which range from an eight-hour day down to "adequate washing facilities . . . where the nature of the work makes it desirable."

Such Reports as these make it impossible to look for any great social betterment through the Ministry of Reconstruction. After reading through the recommendations we are gradually awakened to the fact that these Committee men, with the Master of Balliol at their head, cannot attempt to cure the industrial wound, but merely to keep it clean, and therefore to keep it open. They dare not utter words against the spirit of their masters from whom they have received the commission and the financial means for the Report. They are not just earnest social thinkers speaking what they know at the peril of their life or their livelihood. They are "cogs in the machine," with no movement or voice of their own.

Facts for Whitehall

I WILL SET DOWN SOME FACTS. I WILL ALSO SUGGEST SOME interesting thinking which the facts might occasion.

I

JOHN HORACE BRACEBROW is the well-bred, well-reared, well-educated, well-mannered, well-meaning son of a rich woollen manufacturer at Dewsbury. His taste for early Italian literature, painting, and sculpture has taken him to Florence and the Northern Italian cities. In a letter to his mother–for he is a devoted son -he writes. "Even here at P— there is no quiet. It is like the rest of the Italian towns. All night long the trams run and motors rush . . . . The noise is incessant!"

Has Florence and P— dimmed the memory of Dewsbury? Is noise sacramental, i.e., significant? If so–significant of what? Moreover, is not modern Italy progressive? If so–why not?

II

I WAS TALKING with Josias Dunlop, Member of Parliament for Belfort. I said: "My uncles were well-to-do farmers on under sixty acres." He answered–for it was an answer to my challenge: "I knew a farmer in Fermanagh who gave his children a good education on twelve acres."

This is a fact difficult to pigeon-hole. It is for this reason that in the title of this article I substituted the general locality

Whitehall for the concrete Government Office of the Board of Agriculture. Perhaps some statistician could tell me whether the fact of the twelve acres would be welcomed more by the Board of Education than by the Board of Agriculture. N.B.— I am not quite sure what Josias Dunlop meant by "education."

III

I TALKED TO an old farmer in Somersetshire, As we stood looking up at the hill-sides covered with purple heather, I said: "Doesn't the sun bring out the purple of the heather?" He said: "Aye, thack it do. But when I wor a lad, 'twasn't 'eather as ye'd see thur–t'was wheat. An' the sheep wor a-grazin' in that 'eer gorse, where only rabbuts bide now."

Heather, where once wheat grew; and rabbit runs where sheep once pastured. This seems to go beyond steam-engines and reinforced-concrete roads as a proof of progress.

Stated as a problem it could be put thus: Find out the psychological [if not logical] nexus between this fact vouched by the farmer and the ordinary retort of the average economist, "But where would you be without machinery? Could anyone but a fool think that England could support its population on the land?" Special marks will be given for attempting to state the answer as a "complex."

Compare the economic value of sheep and rabbits; and the nutritive value of wheat and heather. Show by a diagram [or as a graph] how it pays England to grow heather rather than whcat.

IV

I TALKED TO a young town-bred gentleman who loved England so foolishly that he had "left" London for mid-Suffolk. "You have now learned, or are learning, that when you are on the land you need never lose a moment of time or an ounce even of rubbish." "You are right," he said meditatively, "save for tins. Once I bought a number of tinned [or is it canned?] goods. When I and my fellow farmer had eaten them empty, we did not know where to put or what to do with the empty tins. They littered the house for a month. One day, when the rain was pelting, I felt I must act with boldness and promptness. I went out into the rain, dug a deep hole in the paddock, and buried the tins."

This fact throws light on His Majesty the King's speech on "littering" Kenwood.

An engineering expert says it throws light on the mental inferiority of our agricultural population, seeing that a careful treatment of bi-sulphohydrate and radium would yield distyl bi-carbonate which, as he gravely observed, is the active agent of margarine! This proves that the burial-treatment of empty tins is a sign positive of the sub-normal agricultural intelligence.

V

My young man also told me of the local Flower Show. Lady Beatrice Redmeadow had offered a prize for the best loaf of bread baked at home from wheat grown on the farm.

He said: "There should have been a hundred and fifty entries; but there were only five." Within a week the local baker had 'ferreted out" [this was his own rural phrase] the giver of the prize, the winner of the prize, and the other competitors for the prize. With tears in his eyes he had implored them all not to encourage home-baking as it would mean the ruin of his trade.

From this we can easily see how proud we ought to be when we are called a nation of shop-keepers.

Political economists may also learn from this fact how vital trading is–for a trading community; and how essential it is for a nation of shop-keepers to keep shops. To forget this is to arrest progress. Not in bread only–and especially not in home-made bread only–does a nation live. First place must be given to those labour-saving devices such as central bakeries and large breweries which enable people to do for others what those others could do for themselves.

Educationalists may reflect with some justifiable satisfaction that, in the passing away of the old order into the better new, they have kept what was best in such folk [action] songs as:

I can bake and I can brew.
I can make a pie or stew.

VI

Eliza Constant, who is now seeking the worship of God and self-expression on a small holding in Devon, said: "We do our cooking on an oil-stove." Having had some years' experience with a wood fire, I said: "Have you tried to cook with the wood fire? " She answered: "No." I answered: "some novices camping on the north-west coast of Wales kept the feast of the Assumption of God's Mother by cooking a seven-course meal on a fire made of gorse." E. C. then observed: "Oil is our greatest expense; and anxiety. We have to fetch it from the cross-roads every Tuesday. If we forget–we are lost!"

No one who reads this can refuse his praise for those patriotic and financially minded patriots who see that large sums of money are spent in securing foreign oil-fields, rather than in the sordid craft of supporting English wheat-fields.

Oil is very valuable for many purposes.

(*a*) It can be burned as fuel. Here is its incontestable superiority to wheat.

(*b*) It can easily be transported. This, by enabling and inviting many people to move about, leads to great stability in the Commonwealth.

(*c*) By providing large dividends to men born out of these islands it breaks down the isolation so characteristic of our race.

(*d*) By depending on our incomparable Army and Navy it carries the flag where even the name of Alfred the Great or Shakespeare or any of the pre-oil men never reached.

(*e*) England must at once remember and forget the great saying of a recent President of the British Association: "The secret of England's greatness is – coal!" We must remember the spirit of this wise remark. But we must forget "coal" and substitute "oil." Otherwise we shall cease to be great.

VII

I, THE WRITER or rather the chronicler *of these true facts*, will add another fact. Again and again whilst on a railway journey through this, perhaps, richest land in the world, I have played my sad game of counting the sturdy English folk working on English soil. I have passed through as many as a hundred miles without seeing fifty acres that could not yield heavily under the plough or spade. But even in the early days of summer or in the harvest days of August, I have never been able to count a hundred Englishmen or English women at work on English soil.

On thinking over this fact I have no problems to offer, nor irony to veil; but only anger to stifle and tears to dry.

Is Patriotism Dead?

The Quest for Gold and Personal Gain

I CAN ONLY ASK THIS QUESTION. MY READERS MUST ANSWER it. But ever since a talk I recently had with a man of affairs I feel that I must ask the question, even if no answer is forthcoming.

The man came to me, like Nicodemus, by night. In introducing himself he told me he had a training in economic science, and was a director of several important undertakings. Both his study and his experience had driven him to see that the key to the economic difficulty was, not the town, but the countryside. At not a little inconvenience he had come to me to discuss the problem of the land.

We were soon in the sphere of principles, where alone discussion is possible. Now and then he made me repeat a principle that was not at once self-evident or even evident. After a few moments I enunciated a principle which he had no difficulty in accepting, although for a moment he had a difficulty in understanding. I had said quietly but emphatically: "No true citizen should willingly grow richer by what makes his country grow poorer."

When he had finally understood and accepted the principle, he said: "That would be of no account to the men I know."

I was almost stunned by his evident certainty. As his experience with the businessmen of the time was much wider than mine, I had no data for countering his quite emphatic statement. But I think I said something to this effect: "If these men prefer their own interests to the interests of their country, why are they entrusted with the interests of their country?"

* * * * *

It was the war that first brought my mind to realize this and many another forgotten principle of national life. Stated in economic terms, the Great War was a time when this country, like the other belligerents, was using capital as income. For the duration of the war this lavish expenditure of capital gave a show of wealth. Many men, whom war literature promptly dubbed "profiteers," succeeded in snatching large fortunes from this flood of released capital.

It was hardly possible for the men at home not to go after gold when their refusal was only their neighbour's or their rival's opportunity. And there was a further excuse of their action to be found in the fact that even civic, hereditary honours were given to some who were poor at the war's opening and rich at its close.

Yet if it seemed almost excusable to gather wealth in the national squander of wealth, it seemed inexcusable to call this wealth-gathering by the name of patriotism.

But this principle of patriotic self-control which was so obvious during the time of war, was as applicable, if not as obvious, in times of peace.

An example came to hand at once in the fundamental national industry of agriculture. It was clear that if a land-worker had a large farm he could make a large income even when the average yield was not large. Thus a farmer with ten thousand acres could have an income of £10,000 a year if his farm yielded only £1 an acre. In other words, to get £1 an acre out of 10,000 acres would pay the English farmer; *but it would not pay England*! Smaller holdings worked by farmers who owned their holdings could get ten, twenty, or fifty pounds an acre! From this it is clear that a group of persons who aim explicitly at their own wealth, wholly careless of their country's resultant poverty, have forfeited their claim to be accounted patriots or even citizens.

I can never forget the dramatic words of the late Mr. Mackenzie,[19] Director of Cambridge University farm. He pointed out how vast fortunes were made in the United States by the reckless cutting down of forests and over-cropping of virgin soil. At the end of a span of years the United States was the poorer by the (capital) loss of its forest and the impoverishment of its soil; yet some men calling themselves patriotic Americans were vastly rich. With Scottish incisiveness he would add: "God will surely punish this sin against the soil."

I have neither the means nor the heart to investigate the truth of my interviewer's statement. England must be indeed sickening unto death if it is true that the average Englishman does not mind if England becomes poorer by his becoming richer. This picture of a country dying through the selfishness of its citizens is so gruesome that the mind refuses to accept it as a truth. But the mind will equally refuse to accept as patriotism any social scheme which aims mainly at the wealth of the individual and is heedless of the commonwealth.

The Sins of Avarice

A CORRESPONDENT'S LETTER ON "QUESTIONS ON PROFITEERING," contains such subtle economic errors that we cannot let it pass unchallenged. To quote from it:

> What is profiteering? Is it the making of a large total profit, such as the Coats Company has lately declared? Does this necessarily imply any excessive charge for the article sold or any injury to the consumer? May not larger total profits be due to a large trade as well as to a high price?
>
> . . . If two men start a bread shop at either end of a street and sell at a profit of ½d. on a 5d. loaf; and if one gets sale for 200 loaves a day, which gives him a wage of 8s. 4d., his profit and income are moderate enough. But if the trade of his rival, by reason of the position of his shop or his more popular personality, grows to 1,000 loaves a day, is he profiteering because, by charging the same rate of profit he earns £2 a day? Is he wronging his customers?"

Rightly to judge of this clever and characteristic statement of the modern defence of wealth-getting we must contrast it with the doctrine of St. Thomas Aquinas on Avarice. In answer to the question, "Is Avarice a sin?" St. Thomas says:

> External goods have the character of means useful for an end. Hence man's good in them must consist in a certain measure of them; that is, a man must seek to have external riches only in a certain measure, insofar as they are necessary for him in his state of life. In any excess of

> the measure there will be sin; it is evil if he should wish to get or keep them beyond a right measure. This would be avarice, which is defined as "the immoderate love of having."
>
> . . . Avarice can be immoderate in external goods in two ways. First, directly in the getting or keeping of these goods, by getting or keeping them more than he should. This is directly a sin against our neighbour, because external goods cannot simultaneously be possessed by many, and therefore if one man has more than be ought, others have less than they ought.
>
> Secondly, avarice can imply an immoderateness in the internal affection we have for riches, namely, by immoderately loving, desiring, or delighting in them . . . . Consequently it is a sin against God" (IIa, IIae, Q. cxviii, Art. 1).

1. This wise and luminous doctrine of St. Thomas throws light on the question of profiteering mooted by a correspondent in the *Catholic Times*. The writer considered only one point: "Is the profiteer wronging his customers?"

Yet even if we answered this question by a "No," we should have to ask the further question: Is the profiteer sinning against God by avarice, that is, by an inordinate love of wealth?"

A man cannot desire wealth without end; but only such wealth as is measured by his state of life. *Now, this state of life is not determined by his wealth; but the wealth he may lawfully have is determined by his state of life.* The mere seeking to have more and more wealth–and especially the seeking to have more and more wealth by mere investments without any work–is the sin of avarice. Therefore, even if this baker mentioned by our questioner did not wrong any of his customers, he could not go on extending his trade indefinitely for the mere love of riches.

2. Again, it is not so plain that, in the case supposed, he would not be robbing his customers. There is in Ethics what is known as the "Just Price," meaning also the "Just Profit." This is largely, if not mainly, determined by the time, toil, etc., which the seller has expended on what he sells. In the end this

is determined by the right which this seller has to the means necessary for his state of life. It is quite clear that if in selling many things he can reach this state of life by making on each article less than if he sold few things, then to sell many things with the same profit as to sell few would be to exceed the Just Price. Thus, if in selling 1,000 things he can reach his state of life only by making a profit of 2d. on each, then to make a profit of 2d. each on 200,000 is to exceed the Just Price. Now, to charge consumers more than the Just Price is to wrong the consumer.

3. But our correspondent has left out of his question about a seller and consumers a third class, namely, *the other sellers*. Even if the seller of many things was not sinning against God by avarice or (*ex hypothesi*) against the buyers by more than a normal price, we should have to ask: "Is this seller, by merely pushing his trade, wronging other sellers?"

Here the principle of St. Thomas is of poignant application. There are only so many goods, e.g. loaves, to be sold. If one seller, by his industry contrives to sell very many of these goods, the other sellers must sell very few. In other words, they are deprived of their livelihood. A few days ago I was speaking to a cleric whose father, a respectable chemist, had to close his shop through the competition of a well-known multiple chemist firm. In the end this respectable shop-keeper had to take a place as a "hand."

In the three professions, the clerical, legal, medical, there still survives the custom of not intruding in the locality of another professional man. Something akin to this used to obtain amongst the more gentlemanly of the old traders and merchants. But it has long since gone, with the divine sanctions which were at the basis of it. Yet a species of profiteering which would drive the other sellers from their fair profits would be a sin against justice; and therefore against the Commonweal of the Commonwealth.

Memento Mei

("Jimmy" White's Last Message)

FOR A MOMENT THE "HUMAN CAULDRON OF GREED, LUST, and power" slows in its seething as one frank victim passes by his own act beyond its power. The name of James White has a little eddy of influence even after his death; before death swallows up even his name.

It is not his name or even his life, but only his last printed message on Life that we would recall. Death and the official verdict "Insane" make all criticism of his mode of dying almost as gruesome and revolting as the rifling of a grave.

His last printed message deserved to be printed because, once printed and read, it deserves not to be forgotten. It is the message of a man who went up into success like a rocket and then, an hour before his last failure, told his contemporaries of the folly of success.

To those of us who have dwelt under the roof of cloister wisdom the following poignant words of this last message seem an unbelievable form of folly. Even in transcribing them we wonder if they are the words of a mind concerned with or capable of truth, or the truth-like creation of a master of fiction. Let me set its message down as a litany or a dirge.

> I have entertained royalty.
> I have called Dukes and Earls by their pet names.
> I have been on the inside of politics.

I have owned a yacht.

I have run a large racing stud.

I have owned a theatre.

I have had interests in newspapers.

I have brought off some of the largest financial deals.

I have raised over £150,000,000 (one hundred and fifty million pounds) for various undertakings.

I have promoted prize-fights.

I have subsidized boxers.

I have given large sums to charity.

I have made over £750,000 in one day.

I have been feted by all, and called Jimmy White by a world of people.

* * * * *

"From this it must be agreed that I am entitled to an opinion on life."

Too late came the poor beggared soul's verdict on the emptiness of what so many still account the fullness of life. It is tragic that the mind suddenly snapped when it was reaching the essential wisdom of these words.

Judging by the above, it does not "look as if one misses much by sleeping for ever."

These final words, discovered an hour before death, might and should have been discovered or learned whilst the man stood on the threshold of what is called life.

Even the books of the riches-loving Hebrew people, if read, might have given the soul in search of life the wisdom it learned only on the brink of death. Hear some of its essential wisdom.

> Nothing is more wicked than a covetous man.
>
> Why are earth and ashes proud?
>
> There is not a more wicked thing than to love money. For such a one setteth even his own soul to sale. Because such a one whilst he liveth hath cast away his heart.
>
> All power is of short life.
>
> (Ecclus. X, 9-11).

A thousand proverbs as terse and true as this wisdom of the Preacher are still ignored in a world flooded with the output of the rapid printing press.

And a Greater than the Preacher is still giving to unheeding ears his warning:

> Thou fool, this night do they require thy soul of thee. And where shall those things be which thou hast provided? So is he that layeth up treasure for himself; and is not *rich towards God . . . . woe to you that are rich.*

These woes and warnings of a Greater than the Prophets are but the divine authentication of his forerunners unfailing and, alas! unavailing prophecy.

* * * * *

It would be sheer folly to overlook one phrase of the "last article" we are recalling.

> My head swims. My heart throbs and I bend my knees and look to God, for I have been guilty of the folly of gambling, and the price has to be paid.

To God has he appealed; and to God, most merciful, has he gone. But before the world of today goes to God for judgement, let it enter into judgement with itself by putting as a question what the dead man has put as a last despairing confession:

> Are we gambling in life–gambling with life? And if our life is a gamble what final price must be paid?

Let our readers be helped in their judgement by a contribution to those articles on "Life" which the dead man intended to contribute. Mr. W. R. Morris,[20] of Motor-Car fame, writes:

> I look upon life essentially as a practical man. Frankly, I have no use for philosophies. Theories of any sort leave me cold. What knowledge, beliefs, and feelings I have are the result of experience–I mistrust any other sort.

> Life is a practical job. Thinking about it may help some people, but the longer I live the more certain I become that a man can only find himself in the work he does.

Are not these the words of a gambler with life?

There was One who said: "This is the WORK OF GOD, that you believe in him whom he sent" (Jo. vi, 29). The One who said these words was the Way, the Truth, and the Life. All discordant words, even of a successful financier or motor manufacturer, are a defection from the way, a denial of the truth, a gamble with life. If a nation hearkens to them and not to Him its end is with Judas or the swine.

Dear Mother Earth

The Editor of *Columba*[21] by inviting me to speak to his constituency on "The Church and the Land," has laid on my shoulders a responsibility which I will at once share with my readers by offering them the phrase I have set at the head of this article. Just as those who say "Alma Mater Ecclesia"–Dear Mother Church–have made that act of faith which reassures the priest in the pulpit, so, if my readers will but admit the phrase "Alma Mater Terra"–Dear Mother Earth–I shall be encouraged to speak. But if they will not admit the phrase–their back-to-back houses, their canned goods, their margarine, their dole, their tenement bugs and lice be on their own head! I have delivered my soul!

Let me further reassure those who would listen to the truth, even from the present writer. He can solemnly aver that he never was, is not, nor ever will be a politician. And if he speaks straight about the land it is because by a miracle worthy of Pentecost all political parties are now agreed that political salvation is not to be reached except by some return to a land organization. For the moment I am not concerned to prove that this profession of faith in the land is a deep conviction of the heart or is a mere catch-vote slogan of the lips. But the profession of faith is now common to all the political parties at present clamouring for my readers' votes.

This is reassuring. But I may offer my readers still further reassurance, if any is needed. The present writer, not as a politician nor yet as an economist, but as a priest concerned primarily for the salvation of souls, has gradually been forced to admit that he must do his best to change those city conditions which are now so overwhelming as to make race suicide seem the only way out. Indeed, as a priest he must suggest some other way out, or his craft of priesthood will be discredited for denouncing sins which he himself is fostering.

His own observations, which were so disquieting that he hardly dared to turn them over in his mind, are now confirmed by the observations of fellow priests across the seas in the freedom-loving and freedom-giving States of America. Quiet students of the Catholic Church in the United States estimate that the crowding of Catholics–largely Irish Catholics–into the cities of America for the last hundred years has meant the loss of some ten or fifteen million Catholics to the faith!

We must therefore make an Exodus from the flesh-pots of Egypt, or we shall die. Already the state of things is such a denial of the necessaries of human existence that men are beginning to talk frankly of a revolution? But the revolution we need is to leave Egypt and not to assassinate Pharaoh!

There now seems no hope that the modern urban organization of society will give the minimum necessary for a home and a family. All parties seem implicitly to agree on this fundamental fact, because all parties, and even a large growing section of the Labour Party, see in Neo-Malthusian Birth Control, a necessary factor in social reconstruction.

But if town organization when made dominant has proved itself on the one hand equal to giving us Board schools, museums, parks, tram cars, and on the other hand unequal to giving us homes, it has proved its bankruptcy. If the machinery in a stocking factory became capable of making a smoke or a noise, and incapable of making stockings, it would be scrapped!

Now, the main means of scrapping the antiquated social machinery which has produced the slums and tenements of our cities is to leave the city for the country.

But it will be said, "People don't want to go on the land. Would you compel them?"

The reply to this very pertinent objection may open an important furrow of thought. The human race as such has certain duties, some of which are individual and some collective. Thus it is the duty of every individual not to steal—and, indeed, to earn his food by work. But it is the collective, not the individual, duty of mankind to marry and beget children. So necessary is this duty that theologians have thought that if the human race were perishing through lack of marriages, it would be necessary for monks and nuns to enter the married state. Nevertheless, these straits have never been reached in the past, nor is there a likelihood of their being reached in the future. The collective duty of marriage is quite easily fulfilled by those whose inclinations lead them to marry. And in these days it is not the internal inclination to marry that is failing, it is the possession of the external circumstances necessary for married family life.

In the same way the human race has the collective duty of tilling the soil. Every other duty is secondary to this. Just as an individual cannot allow himself to die of starvation, but must live and therefore must work to have a livelihood, so, too, the community must live by tilling the soil in order to have the means of living. And if there was a danger that the human race might starve through lack of land-workers it would be necessary for monks and nuns and clerics to till the soil.

But the human race, God's perfect workmanship, is happily so perfect an organization that this collective duty of tilling the soil is easily met, in normal circumstances, by those who wish to till the soil, just as the collective duty of marriage is fulfilled by those who wish to many.

This desire to till the soil being widespread, there is no need to compel people to go out on the land.[1] Indeed, something like compulsion, or at least temptation, is needed to coax them or warn them off the land. This is the inwardness of the present coal crisis. Coal was, and is, exhaustible wealth. To live

1. Since writing this the writer has found the doctrine in the *Summa* of St. Thomas! (*Suppl.* 41, 2.)

on coal was to live on capital. When England and Scotland elected to live on coal they coaxed labour from the land into the pits by a high money wage. It was criminal and unpatriotic. But we are feeling the after-pangs of the national drunkenness which drew men from the land by the temptation of high wages–i.e., of heavy helpings of token wealth.

We may, then, formulate a historic fact so absolute as to have the character of a law of nature: *The demand for small holdings is so great that it has never been met.* To the present writer, this fact, after the fact of the Catholic Church, is the most hopeful sign he finds in the world of today.

There is, then, no need of compelling anyone to go to the land. The horrible conditions of modern town life are compulsion enough. So little is it necessary to entice people to go to the land that we have an elaborate set of laws making access to the land very difficult. These land-locking laws should be repealed!

But even with these laws operating it is not impossible to facilitate an exodus. It is the opinion of the present writer that probably in no civilized country of the world can land of equal fertility be bought so cheaply as in this country. Nowhere else is agricultural land of so little *real* value as distinct from *token value*. It is still possible to get good freehold land for from £10 to £20 an acre! And one of the Black Houses (which were more sanitary than the mid-Victorian villa) could be put up for £80!

Let my readers consider what would be the present state of Scotland if the powerful and wealthy miners, shipbuilders, ironworkers, had invested one-half or one-quarter of their funds in buying land in Scotland, instead of banking their money or investing it in Government securities, which can be confiscated at a moment's notice!

The Durham and Northumberland Miners' Association have had the wit to build a number of homes for the old miners. It is probable that, in the present crisis, these homes are the most valuable asset of the miners. But what would have been the position if these miners of Northumberland and Durham had bought some of the rich land of Northumberland and Durham?

Even at the rate of £50 an acre they could have had an estate of 4000 acres for £200,000. If that estate had been laid down for food-growing the present coal crisis could not have come to pass.

Even now it would be worth their while setting aside a yearly share of their income for buying land, if only for the purpose of gradually adjusting the community to the change which must come with the exhaustion of the mines. Land in Scotland is cheap enough to prevent this from being a speculative investment.

But, indeed, no investment is so little speculative as investment in agricultural land. In no other kind of commodity does the ordinary buyer run so few risks in buying and in re-selling what he has bought. It is comparatively easy for even an amateur townsman to find out the rent of the land. This can be taken as a basis of value. Twenty to thirty years' purchase will not put the buyer in inextricable difficulties. Moreover, if after a few years possession the buyer wishes to re-sell, he runs less risk of selling at a dead loss than if he were selling any other commodity. All this would go to induce even townsfolk to buy some agricultural land–now that it can be had cheaper than almost any primary commodity in the commonwealth.

But as the buyer of this land is flying from the economic and moral evils brought in by modern industrialism with its machinery and money, he must make it a point of wisdom not to attempt the development of the land with those industrial methods which have proved to be the ruin of the town. He must fly as far as possible from everything that comes to him from machinery and mass production. He must try to do with the minimum of markets. He must not seek as an ideal to have as large a farm as possible; but, rather, to have as small a farm as possible. The craze for the big farm ends by impoverishing the farmer and the farm. Nowhere–and certainly nowhere less than in farming–is mass-production the same as intensive production. Cattle ranches of a thousand acres are almost a criminal impoverishment of the land.

If only the young men about the town whose sporting instincts are satisfied with football or cricket could seek to satisfy these instincts by colonizing England and Scotland, they would find the game of their life. What a fine game it would be to take a wood-axe, an adze, a saw, and hammer and nails; and from the wood on the neighbouring hill build another of these timber-framed houses which have stood for two or three centuries. The present writer speaks with some feeling on this matter, for a wood-axe was almost his daily companion during the four years of the war. It was then he realized that if a man will only go and cut with his own axe he might buy enough timber to build his future wife a cottage for £10! Indeed, an experienced architect to whom I unfolded the romantic game of building one's own house with the materials of the spot agreed that the whole house could be done for £150. He further agreed that he and I could build such a house in three months, and that when it was built it would be more commodious and sanitary than the rooms occupied by seventy-five per cent of the people of my parish in London. Indeed, he agreed with all this so enthusiastically that, in spite of his three score and ten years, he pressed me to spend the next three months building such a house! Alas! I was not master of my own time. Moreover, my job was not to build houses but to help the saving of souls by talking about the building of houses and the building up of homes. But are any of my readers game for such an adventure?

It is not without a sense of hope that, at the Editor's invitation, I have plied my craft of talking to the men who have taken Columba[22] as their leader. This great man had no timber yards or steel works to fall back upon when he began his work at Iona. He had only Iona. But Iona was enough for a man who loved God and his country and had a pair of hands. With the clay and stones under his feet, and the timber from the nearest forest, he began the great game of winning a people to God. He had few tools and little material. Yet his work is still standing after some fourteen hundred years.

St. Columba, pray for us,
Carpenter of Nazareth, help us – to help ourselves.

Towards Hope

SOME TWO YEARS AGO I WAS CALLED TO SEE A STRANGER in the guest-room of our presbytery. He gave no name. His first words to me showed me a gentleman in speech and manner. But these same guest rooms have too often heard the clever fictions of adventurers that I should be instantly caught by the manner and word of a gentleman.

What I expected him to say, he said. I had heard tales like his almost daily for the forty yens of my priesthood. "I have come to ask if you could help me to any job. I have a wife and children." He seemed a man still on this side of thirty. "I would take any job. I would"–here I fancy his voice developed a crack–"I would even take a job as train or bus conductor."

For the moment I was stunned to an icy stillness which must have seemed to him a mean substitute for Christian charity. Floods of thought swept through my stunned mind. I thought I could detect the young citizen-officer whose self-sacrifice had been almost the main ingredient in his country's war victory. At any rate, I saw the Catholic husband and father willing to accept what some would call the shame of menial work, for the love of his wife and children.

Perhaps he did not realize how near I was to an emotional collapse, which I tried to hide under a mask of ice. I could see this gentleman and his gentle wife and their little ones huddled together in one or two of the slum-rooms of our neighbourhood.

But I stemmed my emotion by putting first things first in my wonted question: "Have you ever thought of going out of the town and into the country?" I could see, or thought I could see, a look of wearied disillusionment in his face as he answered: "It doesn't pay. I have land of my own. I have lost money by it."

I could have told him the history of his loss, But I asked: "How much land have you?" I expected him to say: "A house and a few acres in Suburbia." I was amazed when he answered dismally as if naming a heavy debt: "Four hundred acres."

Again I thought that no doubt these four hundred acres which were not able to keep six people, were in one of the wild hill counties of England. I was still more amazed when he answered no less dismally: "In . . . ," and he named one of the most fertile counties in these fertile islands.

Then in sheer desperation, and not in hope, I poured into his seemingly heedless ears a torrent of advice: "You sub-let your four hundred acres?" He nodded assent. "If you sold three hundred and fifty acres and kept only fifty, you might say 'good day" to the job of tram conductor. But to make your fifty acres do what your four hundred acres did not do, you must give up doing what you have done. Instead of seeking first a money return, you must seek first a real return. You must try to live not *by* your fifty acres but *on* your fifty acres. Even the worst labourer's cottage on your four hundred acres would give you twice as much house-room for the family as you hope to have in London, and the soil will give you and your family enough to eat."

Again he nodded assent. But he seemed unmoved as I poured into his ear the first principles of self-sufficiency which alone make a steady foundation for the life of the family and the Commonwealth.

He went away. I could not think I had been of much use to him. I could even imagine and pardon him saying to himself: "I might as well have saved myself the trouble of the walk. If that is all that Catholic charity can do for a young husband and father, I might as well . . . " And I could imagine him trying to forget that he had ever been fool enough to seek my help.

But even if I had tried, I could not forget the young husband and father with his manner and speech of a gentleman and his four hundred acres of good land in the county of . . . and his heroic resolve to take any job, even that of a tram or bus conductor. He haunted my thoughts by day and night. He would push himself with undeniable insistence into speeches at noisy meetings, and when I had speech in silence with Jesus of Nazareth.

* * * * *

A few days ago a man came to see me on an errand of brotherly love for a friend. He introduced himself by saying: "I am a kinsman of the man who came to you and had four hundred acres in . . . "

It was as a voice from the dead. I was alert in an instant. "Well!" It was all I could say in my dismal foreboding. He went on: "Well, he took your advice. All his friends said he was a fool. Even the farmers said he was a fool. But he did what you said."

"Well?" I gasped.

"He did what you said; and he has succeeded."

This is God's truth.

Nature's Call to Work and Thrift

THERE ARE TWO KINDS OF CONSTRUCTIVE POVERTY—(1) the poverty of work; and (2) the poverty of thrift.

The poverty of work is found in the sphere of production; the poverty of thrift is found in the sphere of consumption. By poverty of work is meant the effective will to produce as much as possible of real wealth. By the poverty of thrift is meant the effective will to consume as little as possible of real wealth.

These two forms of poverty are so fundamental that we can discover them in nature; and formulate them in their ultimate form.

When we deal directly with the land we need never waste (1) a moment of time, or (2) an ounce of material.

We do not count as waste of time the hours spent in sleep or the negligible fractions of time spent in going to and from the field of work. As land-work is most efficient when the worker lives, as he usually lives, in the midst of his work, there are few cases where such waste of time need find a place in an economic balance-sheet.

But it is far otherwise in town-organized labour. Indeed the larger the town the greater the waste of workers' time going to and from work. The vast volume of daily passenger traffic in our larger cities may almost wholly be classified as waste of time. Few but the small craftsmen and tradesmen live over their work or crafts. In London some millions spend no little part of each day going to and from their work, and increase this uneconomic spending by the money which they have to pay to the traffic companies.

Moreover, none of the town activities produce primary real wealth. Towns cannot produce food, nor clothing, nor the wood, stones, and bricks needed for housing. Townsfolk are engaged, at their best, in producing secondary real wealth; at their worst, they are engaged in manipulating token-wealth, so as to possess and control real wealth.

Moreover, as the human body needs a certain minimum of fresh air and exercise, the average town-dweller has to spend time in taking air and exercise whilst producing nothing for the wealth of his country.

Contrast this condition of the town with the condition of the country. The smaller the holding the more efficient it is as a unit of production. An efficient homestead means that the worker is never more than a few minutes from his work. Indeed, much of his work is under his home-roof, Nor does he waste any economic values in hiring vehicles to take him to and fro.

Moreover, his work can be always directly or indirectly producing primary real wealth, i.e., food.

Lastly, the leisure required by the town for fresh air and healthy exercise is not needed on the land. So that if a man or woman is so minded they need never lose a moment of time! Of course, in the country, as in the town, human nature being a constant, time may be deliberately wasted. But the contrast between country and town is in this, that, whereas in the country the worker may lose time if he will, in the town the worker must lose time even against his will.

So much for Nature's call to the poverty of work.

The Poverty of Thrift

But her call to the poverty of thrift is no less loud and clear. When once we return to Nature we need not lose an ounce of material.

The factory mass-production of the town is essentially wasteful in material. Although many experts are trying to eliminate this waste of material, no substantial stemming of the waste has yet been found.

But when dealing directly with the land no intelligent worker need waste an ounce of material. Everything that man cannot use for food or clothing or shelter is of value for producing food or clothing or shelter. The present writer was perhaps first directed towards this principle of Nature's poverty of thrift during his three years in Belgium A.D. 1891-1894.

He has never forgotten the impression he received of a people who used everything in Nature to produce further primary wealth. Even the pine-needles and bracken were used as litter; so that the straw might be used as fodder. He used to marvel as he saw the lay brother sacristan carefully throw the dirty floor-scourings about the roots of the trees. Wood ashes serve to make a lexive for washes. Coal-cinders are an excellent metal for paths. Scullery refuse will feed fowls and pigs. Ordure is a most valuable manure.

An important conclusion seems to impose itself. If a town organization is necessarily wasteful, and only a country organization is naturally thriftful, it would seem that a town-organization will inevitably, even if slowly, end with a famine of real wealth, and especially of primary wealth.

Is this the clue to the present world-wide economic crisis?

Absenteeism

(A Shred of Talk with a Seer)

I had accepted his invitation to make my Summer School in his hermitage on the grey Northumbrian coast, where the sun rises daily over the holy Island of Aidan and Cuthbert.[23] During my stay, which now I know to have been shorter than I thought, I had scanty fare for the body. But this scanty fare was seasoned by its recent growth in the soil of Northumberland; and by the banquets of mental fare that followed our silent meals.

I hardly know how began the talk. I will do my best to recall. Our daily speech was wont to begin anywhere; whilst ending mostly in the thought and praise of God. It was a saying of his that "for us men there are infinite beginnings and only one end; but that end infinite."

I found myself asking: "What, then, is this one *Sin of the World* which the Baptist and you make the burden of your dirge?"

Like fork-lightning came the answer: "Absenteeism!"

As I had never heard this word from him before, and as I knew that his contemplations were a growth not only in certitude but in understanding, I looked up with eyes that were a question. Whether in answer to my unspoken question or in vocal meditation on his own answer, he went on:

"Let me kneel down." Here he literally fell on his knees. I could hear the grinding of some pebbles under his heavy fall.

"I shall seem to be a self-appointed judge of men nearer to God than myself. Lest you should think that I sit in judgement on such men or stand to stone them, look on me as kneeling at their feet, unworthy to untie the latchet of their shoes."

His head bowed to the earth. He murmured: "Lord, be merciful to me a sinner."

It was as if the decks of his soul were being cleared for prophetic action. He went on:

> You ask me a question. You have the right of youth to have a reply from your elders. The reply is: Absenteeism is now the *Sin of the World*. It has not grown up in a night, nor has it sprung up only within that world which our language contrasts with the cloisters. Tintern and Kirkstall, and Finchale, if not–[here he turned and bowed his head towards the East]–if not Lindisfarne,[24] are ruins perpetuating the weakness that, like the weakness of Adam, made a garden into a wilderness.

In spite of the prophetic and frightening fire in his words, there was a glisten of tears in his eyes.

> Fields added to fields; lands bartered and bought till the distant holdings of an Abbey were far beyond a day's journey even on horse-back. The Abbey lands, once a sufficient croft around a House of God, now became an estate too wide for the soul of it to be present in every part. Its limbs were swelling. It was sickening to death.

Again his brow was humbled to the earth. I heard him faintly saying–or crooning–"If thou hadst but known the things that were to thy peace."

After a spell he raised himself from his knees. Something like defiance rang in his voice.

> But the new landlords, whose apostasy and avarice had ousted the old, were no longer absenteeism in little. Their countrysides were but farms growing gold to pay a gay life in the train of Tudor or Stuart debauchees. Few of the Abbots had, as few of the new landlords had not, a

> London house which bled a countryside in order to keep itself in being.
>
> These absentees were the prodigals of their day. If we may judge from their knowledge of Holy Writ, they had learned by heart but had never taken to heart the story of their elder brother, whose wisdom came only when he herded and fed with swine. For some two or three centuries they were perhaps the most powerful social institution in Europe. Today they are but a memory or a curse.

The last words, though in themselves so defiant, were spoken almost in an undertone, as if the curse spoken by the lips had no re-echo in the speaker's heart.

A spell of silence seemed to be preparing him for the last act of his swift drama.

I sought to bring him to speaking-point by what was half-question and half-reproach: "The absentee landlords sinned yesterday. I asked you what was the sin of today."

He muttered: "Today was begotten yesterday. If you would be just to the sin of the world you must know its begetting."

I interrupted: "But where is the absenteeism of today?"

At once he became provokingly statistical. With a twinkling eye beating time to a donnish drawl, he said:

> My esteemed friend, have you considered the full significance of the most recent census figures? They tell us that in Great Britain some 75.6 per cent of the population is to be found in urban areas, and some 24.4 per cent in rural areas.[1]

I added sullenly: "Yes, two and two make four."

But he played my bowling with a boundary bit: "In rural areas two and two, if properly planted and manured, will sometimes come to 104."

His donnish drawl passed into a staccato which might have befitted either a philosopher or a prophet:

1. I am writing, *Deo gratias*, in a village far away from the means of making these approximations into accuracies.

> Your town lives on the country; and not the other way about. Your town gallant has to be fed by your country yokel; but your country yokel does not live on your town gallant.
>
> If one man lives in a town, at least one other man has to feed him by living and working in the country. If two men choose to live in the town they will die unless two men choose to feed them by living and working in the country. If a hundred men choose to live in the town, at least twenty men, or perhaps fifty men, have to live and work in the country.
>
> From this we gather the general principle that the greater the number of men who choose to live in the town the greater the number of men who *must*–I say again, *must*–choose to live and work in the country–or the men in the town must die.
>
> From this again we gather the law–and it is a law of death–that a nation which deliberately or indeliberately is beckoning its people into the towns and out of the country–into the streets and out of the fields–is leading its people with *macabre* dancing along the way of death.

He was on his knees once more; and so, too, was I. He moaned:

> Therefore thus saith the Lord. Thy wife shall play the harlot in the city, and thy sons and daughters shall fall by the sword, and thy land shall be measured by a line. And thou shalt die in a polluted land. And Israel shall go into captivity out of their land.
>
> (Amos xi, 17).

* * * * *

When I returned home and found out the sacred book from which he had taken the dirge that had stirred the depths of my soul, I found written at the head of the book: "Amos . . . was called from following the cattle to denounce God's judgements to the people of Israel."

Mass-Production in Agriculture

THE *CHICAGO TRIBUNE*, OF JULY 23RD, HAS THE FOLLOWING item of news: "This year's California peach crop would provide a pack of 17,000,000 cases if it were used, an increase of 4,000,000 cases over last year."

> The pack will be reduced to 9,000,000 cases by agreement.
>
> The Canners' League will buy and destroy 144,000 tons of peaches and will uproot the trees in about 12,000 planted acres.
>
> By this means the growers hope to stabilize the price at a profitable figure.

* * * * *

1. The average man or woman on hearing this destruction of food, says: "How wrong!" As Science, Philosophy, and Ethics rest on the sound judgement of the average person, it is good to seek the cause of this "wrongness" of the Californian canners and peach growers.

2. A cause of this "wrongness" is the confusion between "Mass-production" and "Intensive- production." Even if the Mass-machine-production was Intensive-production, what applies in machinery does not apply to land.

These peach-growing lands do not necessarily yield more peaches per tree or per acre than the peaches that cling about cottages in a land of small holdings. But white sheep give more wool than do black sheep, because there are more of them. So, too, Californian peach acres give more peaches than Spanish or Italian acres, because there are more of them.

Until this principle is seen men and even legislators who propose Agriculture Bills will go on confusing Mass-production with Intensive-production, to the further stagnation of agriculture.

3. Another cause of this "wrongness" is the fallacy of "Markets."

It needs some very controlled thinking to realize the exact position of markets in a people's life. But at least we can say that, even if markets are a necessity, they are never a primary necessity.

Now to treat as a primary necessity what is a necessity but not primary is to throw the whole economic organization in confusion. It can never be sufficiently grasped that in an orderly series, whether of beings or of actions, if one item of the series is out of its order the whole series is out of order. Especially if what is first is not put first, the disorder is complete.

To take an example. Railroad transport is (perhaps) necessary for our modern economic life. But railroad transport is not of primary necessity; because transport itself is not a primary necessity. If the economic consciousness becomes so bewildered as practically to look on railroad transport as primary, all kinds of economic and financial expedients will be needed to help out the mistake; and the country's finances will grow steadily more confused. For example, bounties on home-grown wheat may be in reality bounties to home-railways with their foreign shareholders.

Almost greater confusion would necessarily arise and has unfortunately arisen from a widespread belief in the primary necessity of markets. This confusion is inevitable even when market means exchange; still more inevitable when market means "market exchange." It cannot be too clearly recognized that the economic primaries are but two: Production, Consumption. Other activities, such as Exchange, Distribution, Transport, Market, Price-fixing, Money-value, are never primary even when practically necessary.

Therefore economic schemes based on a practical conviction that these non-primaries are primary will bring economic confusion; and the greater the scheme, the greater the confusion. National schemes based on this mistake will tend to national bankruptcy.

4. "Stabilize the price." To understand this phrase is to unearth another cause of the "wrongness" of destroying peaches–root and fruit–in the interests of a balance-sheet with its profit-and-loss account.

Any attempt to "stabilize prices" is as quixotic as an attempt to stop the tides. Price is the money-token given for a thing. But tokens, as such, have their meaning or worth or value entirely from the human will. As this will is a free-will it can do this or that, can give this meaning or that, can fix this value or that. There is no fixed relation between a money-token (say, a pound sterling) and a thing (say, a ton of peaches). That relation between the token and the reality being a free-will relation, can change as freely as the will. All attempts to "stabilize the price" of things is an attempt to stabilize the free-will; and all hope of stabilizing the price is an economic denial of free-will. It is time we recognized that sound economics cannot be based on unsound psychology.

5. "Profitable figure." Deep thought on the word "Profit" will lead us to detect another cause of the "wrongness" of destroying food.

More and more is the present writer beginning to see that all economic discussion should now be in simple, concrete, ultimate terms, understanded of the people. Words like capital, interest, price, stabilize, rationalize, etc., finally darken counsel. If Plato was right in saying that the true value of poetry could be found by turning it into prose, it will be equally right to say, in the present bankruptcy of things and thoughts, that the true value of an economic statement can be found by turning it into simple, concrete, ultimate words. Let us apply this to the phrase, "By destroying 144,000 tons of peaches and uprooting the peach trees in 12,000 acres, the peach-growers hope to stabilize the price at a profitable figure."

This may be re-stated as (*a*) "The growers hope that the destruction of things (i.e. peaches) will bring them profit."

By the substitution of the word "good" for "profit" this may be further re-stated: (*b*) "The growers think that to destroy things is good!!"

Or again: The growers think that by destroying real-wealth they may obtain possession of money-wealth (or token-wealth).

* * * * *

The proposition (*a*) reminds us of Von Moltke's[25] famous saying: "War is to destroy–and again to destroy." It would therefore seem that an organization which is helped by the destruction, not of what is bad (like a cancer) but of what is good (like a peach) is in its essence, War.

The modern mind is so bewildered by the stir and din around it that it hardly knows the meaning of the word "good." We can use the word "good" by itself only when we mean moral goodness, e.g. "John is good" means "John is morally good." But if we mean that John is a good mathematician, or pianist, or politician, or salesman, or peach-grower, we cannot say, "John is good." We must add: of persons, what they are good at, e.g. John is good at mathematics or peach-growing: of things, what they are good for, e.g., peaches are good for health.

For this reason we cannot say "peach-growers are good by destroying peaches." Even on the economists' showing we should have to add: "Peach growers are good salesmen by destroying peaches." Or again: "To destroy peaches is good–for the peach-growers' wealth."

* * * * *

Take the statement: (*b*) The modern mind has become so over-clouded that it does not see the cure, or even the nature of its ailment. Men of every nation and of every party in every nation are agreeing that the modern world is suffering from an economic want. Experts like the late Conservative Chancellor of the Exchequer (Worthington-Evans) and the late Labour Chancellor of the Exchequer (Mr. Snowdon) call it "the greatest economic blizzard the world has ever seen."

Now this want or lack is either a lack of things, or a lack of tokens. (For the moment we will not consider the Pope's accurate diagnosis that the present economic crisis is not a lack of things or of tokens but an unjust distribution of things.)

If there is a lack of things, it is evident that this lack of things cannot be met by any making or distribution of tokens. For example, a shortage of houses can be met, not by printing a number of coupons with the word HOUSE; but by building houses.

On the other hand, if there is a lack of tokens it is evident that this lack of tokens need not be met by a making of things but merely by the making of tokens. Yet it is ridiculous to consider that the present world blizzard in matters economic is merely a dearth of tokens, which could be met by an issue of paper-money.

Yet again, on the other hand, it is impossible to overlook the fact that men like these Peach Growers' Association are concentrating on the token rather than on the reality. Such is their need or wish for token-wealth they are not only overlooking, but even destroying real wealth. In the long run, such a policy means bankruptcy; if it is a national policy, it means national bankruptcy.

* * * * *

"Nature, to be commanded, must be obeyed," said Francis Bacon.[26] As nature is the work of infinite intelligence; and as men of the highest human genius can give us only a work of finite intelligence it follows that a return from human artificiality to the divine simplicity of nature is alone Progress.

Unfortunately the speed-and-noise-stunned world refuses "to go back" to such simplicity–although it is not a going back but a going on; and although Jesus of Nazareth is the only Redeemer of a lost world.

Group Home-Colonisation

My readers will bear with me giving vent to convictions which seem so vital and urgent that their vital interconnection could be made clear only by forgetting the urgency of the call to action.

So great disaster has been the end of all efforts to "restore agriculture" that almost against our will we have at last discovered the secret of this failure in this phrase "to restore agriculture." The true phrase ought to be "to restore families." Modern efforts to renew the face of the earth have been directed to increasing the number or bulk of things which the land could yield. Statesmen and businessmen have acted together to increase the yield of wheat, or oats, or cows, or pigs, or horses, or tea, or sugar, or bananas, or any other commodity which a seller can sell for a profit. Even today the be-all and end-all of a land holding is to raise some commodity to send for sale in a market.

Yet it should strike political economists that the proper object of political economy is not a thing but a person—and, indeed, that fundamental moral person "a family." All land policies should, therefore, aim not at raising as many commodities as possible from the soil, but of settling as many families as possible on the soil.

If only we could settle the greatest possible number of families on the soil, the commodity question would settle itself. To the average town reader this may seem a dark saying. But it will become a clear saying to the town reader who realizes that the plot of land yielding the most is not the hundred-acre field tilled and harvested by motor; but the small holding or allotment tilled and harvested by hand and foot.

Moreover, as the essential of good land-work is co-operation or group work, the time is come when some effort must be made to settle Catholics on the land *in groups*. Individual land movement is good, but is far from sufficient. It is hardly more than the beginning of a beginning. It is a fine, heartening gesture towards the one thing necessary.

But the invitation to group movements, which some of us now see to be not only necessary but urgent, is commonly met by the statement that people will not or cannot undertake them.

To this many replies may be made, almost every one of which is a complete reply.

To argue that the colonization of rural England is beyond the intelligence of modern Englishmen is to assume a mental regress which every theory of progress must deny. English Catholics in the time of Elizabeth and James were intelligent and dogged enough to form successful group colonies in Maryland, where the difficulties were greater than those which would now face Catholic colonists in the deserted England of today. Therefore, until such time as the Catholics of this country have made the great effort and have thus died by their own hand, the present writer will believe that if only we could find our Lord Baltimore the epic of Maryland would be again enacted in the wilds of England.

But it is amazing what Catholics will do by voluntary effort when the alternative is compulsion and danger to their faith. Now some compulsory land colonization is the only political or social expedient left untried. Already the nation's age-long distaste for tariffs has been discarded or overcome. Yet not even the most sanguine tariff zealot would guarantee that the system would give to England a success which it has failed to give in tariff-ridden United States.

With the partial or complete failure of tariffs to bring national solvency the only untried expedient will be compulsory colonization of the land. Practical politics will not long leave this last expedient untried.

For this reason the present writer believes that at least, in the end, Catholics will rouse their dormant powers of voluntary co-operation to forestall or resist the compulsion of a movement which, to be quite national, would have to be as secular as the non-denominational system of education.

Year by year the fruit harvest and the hop harvest are a lesson-in-little of group movement towards the land. As the fruit season draws near there are parishes in East and South London which witness the organization of parish groups who for some three months leave city life for work on the land.

When the present writer first went out with the fruit-pickers in the Wisbech neighbourhood his heart sickened at the conditions created by mass production applied to the countryside. Something almost worse than the London slums which these fruit-pickers had temporarily quitted awaited them amidst the orchards. The shacks in which they herded were fitly called "bunks." Everything about the massacred countryside was depressing except the men and women and children who had quitted one slum for a worse. Whilst the local country folk and townsfolk looked on these fruit-pickers almost as human vermin, there was one observer who saw in the almost exuberant good spirits of these city dwellers a quality that lifted the depression.

In spite of their boisterousness, and even occasional riotousness, these city dwellers succeeded in doing two difficult things: they kept a very definite family life, under almost impossible circumstances; and they allowed themselves to be organized by one of their own fellow-Catholics into definite groups.

The group should be as far as possible a self-subsisting unit. Of course, the more farmers it contains the greater will be its power of self-subsistence. But it should consist of various other secondary craftsmen, as did Nazareth. It should have its worker in wood, like St. Joseph. It should have its weaver, who will be all the more self-subsisting if he has a lug of land and another craft like the weaver-fishermen of Ireland and Ceylon. It should have its worker in leather to make footwear and harness. It should have its schoolmaster, who will be all the better master of realities if he, too, has another craft like the old schoolmaster at Ballyquentin,[27] who trained up some noted master-mariners whilst making fishing nets.

Such a group has the stability and permanence of co-operation. And in these days when God's call to co-operation is being compulsorily applied in a new Russia that knows not God, we believers in Jesus of Nazareth must be up and at work lest we miss the sun.

Fifteen Things a Distributist May Do

(An Open Letter to J.— B.—, A Young Gentleman With Desires)

YOU SAY YOU ARE AT SEA WITH A PEN; AND THAT ANYHOW, Fleet Street, not to say Little Essex Street, is in no need of recruits. You feel that Distributist modelling and designing has been well and truly done; but that the modellers and designers will have wasted their brains if some simple folk like you don't attempt to carry out the designs. You ask dramatically, "Don't tell me what to think for I subscribe to *G.K.'s*. But tell me what to do."

I will therefore set down fifteen things, any one of which would be good to do. They shall be fifteen for two reasons. Because if I set down the hundred and one things you might do, it would fill a whole issue and not just one article in *G.K.'s*. Fifteen gives a choice such as a man has, say, in choosing a cravat, a livelihood, or a wife.

I will not begin at the beginning. I will begin anywhere and go on anyhow. But, indeed, when things have come to such a state of social untidiness as they are at present, a beginning can be made anywhere and anyhow. The one thing necessary is to begin.

1. If you have a mantelpiece, remove everything from it except perhaps the clock. If you are fortunate enough to have no mantelpiece, remove from the walls of your room all pictures and such like, except a crucifix. This will teach you the Poverty of Thrift. It may be called an empiric approach to Economics.

2. Clean out your own room daily. Clean it if possible on your knees. This will teach you the Poverty of Work. It will also prevent paralysis of the knees. But a paralysis which has reached the knees will soon reach the hands and the brain, if not the tongue.

3. For forty days or more–say, during Lent–do not smoke (and neither grouse about it nor boast about it). This will also improve your eyesight. It will also improve your insight into the tangled economic question: (tobacco) combines and how to smash them.

4. Buy some hand-woven cloth. Wear it. Buy some more. Wear that too. Remember the noble advice on how to eat cucumber: Buy a large cucumber, cut it into two parts (equal or unequal). Eat one part. Then eat the other. Your home-spun will instruct you better than the Declaration of Independence will instruct you on the dignity and rights of man.

5. Buy boots you can walk in. Walk in them. Even if you lessen the income of the General Omnibus Company, or your family doctor; you will discover the human foot. On discovering it, your joy will be as great as if you had invented it. But this joy is the greatest, because no human invention even of Mr. Ford[28] or Mr. Marconi[29] is within a mile of a foot.

6. Find another young Distributist, with or without University education, but with brains and feet.[1] Invite him to use his feet by tramping with you across any English county, say, water-logged Staffordshire during the summer holidays. Invite him to use his brains by standing on his feet, but not on his dignity, in market-places, telling the village-folk what is the matter with Staffordshire. This will lead him to tell them what is the matter with himself.

If you will keep at it for three weeks or a month, your advice on How to Save England will be more valuable, though,

1. Do not believe X.—, who says they are not to be found. The truth is that X.— has lived so long in stunt circles for the last six months that he has become prematurely infantile.

I admit, less valued than that of the entire Board of Directors of the Old Woman of Threadneedle Street.[30]

7. If you fail to find a fellow-tramp, or if you covet the heroism of the dug-outs in a time of peace, spend your summer holiday as a farm-hand. You will not be worth your keep; but it will be worth your while. If Babylondon has not befogged your "intellectus agens"–your active intellect, in the noble phrase of Scholasticism, you will gradually see the Poverty of Work. This is the other empiric approach to Economics.

8. If through the machinations of Beelzebub or his fellow-devil Mammon, your house is in suburbia, plant your garden not with things lovely to see like roses, or sweet to smell like lavender; but good to eat like potatoes or French beans. At the end of two years you will have done three things: (1) You will have a higher appreciation of yokel-intelligence; (2), you will have a wider knowledge of Natural History (especially of slugs and the like); and (3), you will have a sardonic scorn for the economics of our present Sewage System. In other words you will have had the beginnings of a liberal education.

9. I will now approach a matter of your reputation. If you take the advice offered, you may be accounted a fanatic. But fanatic or no fanatic, here is the advice. For twelve months, if possible, or at least for twelve days, do not use anything "canned," neither canned meat nor canned music.

This will throw you back on what is called Home Produce. This in its turn will show you the right expression to put into your singing of "Rule Britannia."

10. I will now appeal to the artist that is within every one of us. Art, as you know, is the right way of making a good thing. There is no right way of making a bad thing. Not only do something, but make something–a cup of tea, a boiled egg, a hat peg (from a fallen branch), a chair!

This heroic attempt to make something will enable your friends to practise their wit by saying you have only made an ass of yourself. In order to hear this gibe stolidly, read up about "the ass's colt at the crossways."

11. Talk your young architect friend into spending two weeks of his holiday making an abode (formerly called a house[1]). He is thinking in terms of Brick-combine bricks, Timber-combine timber, Steel-combine steel, Cement-combine cement, Building materials-combine building materials. Drag him out into England that still grows oak, elm, ash, beech, fir, larch, etc. Give him a wood axe, a hatchet, an adze, and a few tools. Tell him from me that if in two weeks and for less than £100 you and he cannot make an abode more spacious and sanitary than ninety per cent of the dwellings in the Borough of Westminster or St. Pancras, you should be certified. This may be called the Strain Test, enabling you to know whether he has brains enough to be your friend, even if he has brains enough to be an architect.

12. Set down for the information and inspiration of young Distributists one hundred answers to the usually despairing question: "How can I get out of London?" Begin with the simplest answer: "Walk out." You may find that some of your most promising Distributists will walk no more with you. Do not be despondent at this; because it may make your own Distributism more sea-worthy.

13. As you are not yet married, and as marriage is the fundamental state of life as well as the unit of the Commonwealth, make up your mind whether you are called to this state. If you make up your mind to marry, do not marry merely a good wife: marry a good mother to your children. A wife that is a good mother to your children is the Angel of the House; the other sort is the very devil.

14. Before asking her hand and her heart, tell her how to test you. Advise her to ask herself not whether you would make her a good husband, but whether you would make a good father to her and your children. A wife that is not housewife,

1. The word "house" is now becoming obsolete. Collections of flats are not a house. For the moment the genius of the English language seems unequal to the task of giving these collections a name.

and a husband that is not a good house-band are heading for Admiralty Probate and Divorce!

15. If you do not feel called to the state of marriage vows, there is another state of vows–where mysticism and asceticism prove themselves the redemption of Economics.

But–well–God-speed you, as they say in lands of the old culture.

THE END

Notes

1. Hexemeron or Hexaemeron. A term that signifies "six days," though it typically refers to the six days of creation mentioned in Chapter 1 of the Book of Genesis.

2. Ur. A city in Mesopotamia some 8,000 years ago. In the Bible it is referred to as Ur of the Chaldees because of the Chaldeans who settled there around 900 B.C.–forming the Sumerian civilization. Ur was located in modern-day Iraq. Haran was also located in Mesopotamia, and was the home of Abraham's family once it left Ur. It is currently located in southeast Turkey.

3. *Quicumque vult.* Latin for "Whomever wishes." Also a shorthand way of referring to the Athanasian Creed, or Creed of St. Athanasius, since its first line begins with these words: "Quicumque vult salvus esse, ante omnia opus est, ut teneat catholicam fidem (Whosoever will be saved, before all things it is necessary that he hold the Catholic Faith)."

4. Solon (638–558 B.C.). Athenian lawmaker who became known as one of the "seven Sages of Greece." He was appointed Archon of Attica in 594 B.C. in an attempt to quell civil disorder. His ordinances–the *seisachtheia*–worked so well that he was invited to write what became known as the Solonian Constitution. He introduced a set of laws that were a compromise between Democracy and Oligarchy, and which divided the population into four classes based on property ownership. He also introduced trial by jury, established administrative and consultative bodies, and regulated weights and measures.

5. Lycurgus (c. 800 B.C.). Known as the "Father of Sparta," he was made immortal by Plutarch. A man of immense integrity, he sought to bring stability and happiness to Sparta. Along with King Charilaus, he took over Sparta by force and imposed order upon it. Faced with massive inequality in land ownership, he divided it equally amongst the population, insisting that merit, not wealth, determined a man's worth. He banned the ownership of gold and silver, and introduced Spartan coins that had no intrinsic value. His most far-reaching reform was the banning of useless occupations with the result that luxury trades, merchants, prostitution, fortune telling, jewelry production, robbery, and bribery disappeared rapidly. Once his laws were well established, he renounced power and went abroad. His laws remained in force for 500 years in Sparta, and made it the strongest of the Greek city-states.

6. A reference to St. Thomas Aquinas (1225–1274). The official Philosopher of the Catholic Church. His first Summa was the *Summa of Christian Teaching*, which was prepared specially to deal with those who did not have the Catholic faith: pagans, Jews, Greek schismatics, and Muslims. His second, begun in 1266, was the *Summa Theologica*, for which he is most famous, and which was a beginner's (!) introduction to Catholic theology. Mary Clarke, the Thomist writer, says: "To know St. Thomas is to know the medieval mind at its finest, its most powerful, and, indeed, its most modern. For he is timeless and timely, a man for all ages."

7. A reference to the pioneering work of George Wyndham (1863–1913). A liberal Tory, Wyndham was Chief Secretary for Ireland from 1900 to 1905. An ambitious reformer, he is best remembered in Ireland for his success in putting the Land Act of 1903 on the books, whereby Irish tenants could buy out their owners steadily and piecemeal. It was an action that allowed much of the land confiscated from the Irish by English forces to pass back into Irish hands.

8. Lewis Carroll (1832–1898). Pen name of the English Victorian writer Charles Lutwidge Dodgson. He graduated from Oxford in 1854 in Mathematics, and went on to produce *An Elementary Treatise on Determinants* (1867) and *Euclid and His Modern Rivals* (1879). Although he focused on literary works only from 1880 onwards, it remains that he had established his reputation beforehand, producing *Alice in Wonderland* (1865) and *Through the Looking Glass* (1872). His most famous poem was *Jabberwocky*.

9. *Domine! Salva nos: perimus*. Latin for, "Lord, save us; we perish." St. Matt xviii:25.

10. Sir Laming Worthing-Evans (1868–1931). British Conservative Party Secretary of War from 1921 to 1922. He was made Chancellor of the Exchequer in 1928, though it is widely believed that his promotion had been opposed by the City of London.

11. Viscount Philip Snowden of Ickornshaw (1864–1937). Strict English Methodist, Socialist, and teetotaler. Became National Chairman of Independent Labour Party in 1903. Elected to Parliament in 1906. Appointed Chancellor of Exchequer by Ramsey MacDonald in 1924, and again in 1929. His suggestion in 1931, during the Great Depression, that the Unemployment Benefit of workers should be cut so enraged a Labour Party already highly critical of his "un-socialist tendencies" that both he and MacDonald were expelled from the party. A noted speaker, he wrote *A Socialist's Budget* (1907), *Socialism and Teetotalism* (1909), and *The Living Wage* (1909).

12. Anicius Manlius Severinus Boethius (c.480–c.525). An Italian philosopher, poet, Senate politician, and Consul from a high-ranking family. Fluent in Greek and well-versed in Greek philosophy, he is best remembered for his work, *The Consolation of Philosophy*, though it only became widely known and regarded some 300 years after his death. It remained in

vogue almost 1,000 years. He also translated Aristotle's *Categories*.

13. St. Thomas More (1478–1535). Scholar and Chancellor of England during the reign of Henry VIII. He was martyred for his fidelity to the Catholic Church and for his refusal to accept the King as Supreme Head of the Church in England. He wrote the enigmatic novel *Utopia*.

14. The Napoleonic Wars. These take their name from the celebrated Corsican military commander, Napoleon Bonaparte (1769–1815), who was declared Emperor Napoleon I in December 1804. The wars lasted from 1803 to 1815, but were in reality merely a continuation of the conflict that was sparked by the French Revolution of 1789. During these wars, coalitions came into being to fight Bonaparte, and were variously composed of Austria, Great Britain, Russia, Sweden, Prussia and Saxony.

15. A likely reference to the words of Hiram, the King of Tyre in Lebanon who lived during the times of King David (1085–1015 B.C.) and King Solomon (970–928 B.C.), who wrote to Solomon, "we will cut down as many trees out of Libanus, as thou shalt want." 2 Paralipomenon ii:16.

16. Tutankamen. An Egyptian Pharoah who ruled from 1332 to 1323 B.C.. He died in the ninth year of his rule in the war against the Hittites. He was made famous in modern times thanks to the discovery of his riches-laden tomb by Dr. Howard Carter and Lord Carnarvon in 1922. The tomb was wholly intact after millenia–a rarity in itself. The point that McNabb is making is that Tutankamen was buried in three coffins: the outer two coffins were made of wood covered in gold, whilst the inner coffin was of solid gold alone.

17. Probably a reference to a series of eight plates etched and engraved by William Hogarth (1697–1764), which traced the life of one Tom Rakewell. Having inherited his father's fortune, the middle class youth, driven by his aristocratic aspirations, descends by stages into madness. In other words he represents "the whoring after false gods," and since the final plate depicts a demented "Britannia," it *implies* that Hogarth believed that his country had gone made. *A Rake's Progress* has inspired many artists in the twentieth century, including opera composer Stravinsky.

18. This is probably a reference to the poem, *Ichabod*, written by John Greenleaf Whittier (1807–1892) in 1850. The word is Hebrew and means "inglorious"; it is to be found in 1 Kings iv:21 and xiv:3. Perhaps McNabb's meaning is best captured in the third-to-last stanzas of Whittier's poem:

> Of all we loved and honored, naught
> Save power remains;
> A fallen angel's pride of thought,
> Still strong in chains.

19. Kenneth James Joseph Mackenzie (1867–1924). Following his education at Fort Augustus College, the Agricultural College, Wye, the British Dairy Farmers' Institution, Reading, and elsewhere, Mackenzie became professor of agriculture at Wye. In 1907 he joined the staff of the School of Agriculture

at Cambridge University , and became director of the University Farm. He received the honorary degree of M.A. from the University, and in 1914 was elected reader in agriculture. He also served as consultant to the Institute of Animal Nutrition, was a Governor of the Royal Agricultural Society. His publications include the monograph *Cattle and the Future of Beef-Production in England* (1919). According to his co-author, Francis Hugh Adam Marshall, he was "never satisfied with second-hand knowledge, but spent all his available time and much of his available cash in seeking, and, so far as was in his power, in obtaining, a direct personal experience of every phase of agriculture, from bacon factories in Wiltshire and Essex to mountain sheep breeding in Wales and Scotland, and from the small holdings of Belgium and Holland to the extensive farming of Australia. This characteristic, combined with his vivid personality, was probably the main secret of his success." He was also president of the Cambridge University Catholic Association, an active member of the Universities' Catholic Education Board, and, again according to Marshall in *Nature* 113, no. 2851, for June 21, 1924 (896–97), "he took a leading part in the corporate life of the Catholic community in Cambridge, and he ever showed a deep devotion to the faith of his forefathers."

20. William Richard Morris (1877–1963). Although he became famous for his car manufactures, he started out in life making bicycles and motor cycles. He produced his first car, the 1100cc "White and Poppe" Morris Oxford in 1913 at Cowley with financial backing from the Earl of Macclesfield. His most famous car, the Morris Cowley, appeared in 1915. However, he made his real money during WWI when the Ministry of Munitions gave him a contract to manufacture mine sinkers. He became Sir William Morris in 1929, and Lord Nuffield in 1934.

21. *Columba*. Fr. McNabb is most likely referring to the publication of the Knights of St. Columba, a Catholic, fraternal and ritualistic organization founded in 1919 in Glasgow, Scotland, whose aim was, in the words of P.J. O'Callaghan, writing in *Columba* in 1929, to "seize for Catholics those material benefits of which we are often deprived, owing to either national or religious prejudice. The Order would be neither a Confraternity, nor purely a religious society. We would occupy ourselves almost wholly with the things of this world, and what was spiritual in our Order would tend towards reconciling the business of this life with that of the next" ("The Birth of the Order," reprinted in *Columba*, June 2009, 8–9).

22. St. Columba (521–597). A member of the royal clan, O'Donnell, of Donegal in Ireland, he was baptized "Colum"–meaning "dove"–and this was rendered in Latin as "Columba." Studied under St. Finian at Moville, and entered the monastery of Clonard where he became one of the twelve Clonard disciples who became known in history as "The Twelve Apostles of Ireland." After founding many Irish monasteries, he went to Scotland in 563 to spread the Gospel, landing on the island of Iona on May 12, which is located half way between Ireland and the Picts of Caledonia. Some years

later he moved to convert the northern Picts and wrought many miracles. When he was not on mission he resided at Iona and governed many monasteries, both in Ireland and Caledonia, which regarded him as their founder and father. He is said to have written 300 books, of which just two have survived: *The Book of Durrow* and a psalter, *The Cathach.*

23. An obvious reference to England. St. Aidan died in 651. He was consecrated the first bishop of Lindisfarne, England, in 635, although he was an Irish monk who had gone to Iona in 630. In Bede's *History*, St. Aidan is very lavishly praised. St. Cuthbert (c635–687). It is said that he had a vision of St. Aidan in 651 which moved him to become a monk. He became the Prior of Lindisfarne in 664, and its bishop in 685. He was so renowned for miracles that he became known as the "Wonder Worker of England." In 698 his relics were ordered to be transferred, and it was then discovered that his body remained uncorrupted.

24. The ruins that Fr. McNabb's interlocutor refers to are ruins of Cistercian and Benedictine monastery churches and buildings. The Cistercians were a strict line of Benedictine monks that flourished in England before the Reformation, when much of their land and property was confiscated by Henry VIII. Tintern Abbey was the second Cistercian foundation in England, and the first in Wales. Its ruins are located in the Wye valley, in Monmouthshire, about six miles north of Chepstow where the Wye empties into the River Severn, which itself leads West, between Wales and England, into the Bristol Channel. Kirkstall Abbey was one of several important Cistercian foundations in Yorkshire, England, along with Byland, Fountains, Rievaulx, and Roche. The ruins of Kirkstall are today situated on the outskirts of Leeds, about three miles from the center of the city. Finchale Priory was a Benedictine house dependant upon Durham cathedral, which began as a hermitage founded by St. Godric (c1065–1170). Its ruins are located in County Durham, in the North of England, alongside the River Wear. Lindisfarne is an island 2 miles off the Northumberland coast of England and is just southeast of Berwick in Scotland. It is 3 miles long and 1.5 miles wide, and is joined to the mainland at low tide. Lindisfarne monastery was famous as the mother church and religious capital of Northumbria because St. Aidan (*supra*, note 23) founded his See there in 635 at the request of King Oswald. During the eighth and ninth centuries it suffered repeated Viking attacks, such that the bishopric was transferred to Durham. But the priory itself was re-established as a Benedictine institution late in the eleventh century, and its ruins remain on Lindisfarne today.

25. Helmuth Von Moltke (1848–1916). Prussian General whose famous uncle, Moltke the Elder, had gained important military victories in the period 1866 to 1870. Helmuth replaced Alfred von Schlieffen as the German Army Chief of Staff in 1906, though he generally maintained his predecessor's war plans. When war broke out in 1914, he proved wholly indecisive in the invasion of France with the result that the Battle of the Marne produced

immense confusion. The Kaiser, Wilhelm II, therefore, replaced him with Erich Von Falkenhagen.

26. Sir Francis Bacon (1561–1626). A Renaissance courtier who at the age of 12 had gone to Cambridge University. It was there that he turned against Aristotelianism and Scholasticism in favor of Renaissance Humanism. He entered Parliament in 1584 and advanced under King James I, becoming Viscount St. Albans and Lord Chancellor of England in 1618. He liked to see himself as an apostle of a new methodological interpretation of the universe, and outlined his views in *The Advancement of Learning* (1605) and *Novum Organum* (1620)–though other authors have seen the influence of Kabbalism in his work. He is regarded by many as one of the intellectual founders of the modern world.

27. Ballyquentin. Usually rendered "Ballyquintin," it is a small coastal and fishing community located a few miles from Portaferry on the southern most tip of the peninsula of County Down, Ireland.

28. Henry Ford (1863–1947). Member of a prosperous Irish family from Dearborn in Michigan. He became a machinist in Detroit in 1879, and had risen to be the chief engineer of Edison Illuminating Company by 1893. He founded the Ford Motor Company in 1903, making his name with the production of the Ford Model T in 1908. In 1913, he introduced the conveyor belt assembly system, which reduced both costs and the need for skilled workers. Due to its dehumanizing effect on his employees and its consolidation of the factory system, Ford, with this invention, became a symbol for Distributists and other social Catholics of all that was wrong with the modern economy and modern methods of production.

29. Guglielmo Marconi (1874–1937). Italian inventor who produced the first practical system of wireless telegraphy in 1895. In 1896, he traveled to London and was granted the world's first patent for such a system. In 1897, he founded a company to promote his invention; in 1900 it became known as "Marconi's Wireless Telegraph Company." He was awarded the Nobel Prize for Physics in 1909.

30. Old Woman of Threadneedle Street. This is a reference to the Bank of England, located in the street of the same name, based upon a caricature by James Gillray (1756–1815) entitled "POLITICAL-RAVISHMENT,–or–The Old Lady of Threadneedle-Street in danger!" It was produced as a hand-colored etching and published by Hannah Humphrey on May 22, 1797. As described by the New York Public Library, where the etching is held as part of the James Gillray collection (Image ID: ps_prn_cd30_431, in the Humanities and Social Sciences Library), the etching was "another satire on the currency crisis and the substitution of paper money for gold." The Bank of England was personified as the "Old Lady of Threadneedle-Street," and portrayed as the victim of William Pitt's very improper advances. Pitt's hat, in the etching, "partially hides a list of loans, alluding to the Whig accusation that Pitt intended to spend the nation's gold on war."

About IHS Press

IHS Press believes that the key to the restoration of Catholic Society is the recovery and the implementation of the wisdom our Fathers in the Faith possessed so fully less than a century ago. At a time when numerous ideologies were competing for supremacy, these men articulated, with precision and vigor, and without apology or compromise, the only genuine alternative to the then- (and still-) prevailing currents of thought: value-free and yet bureaucratic "progressivism" on the one hand, and the rehashed, *laissez-faire* free-for-all of "conservatism" on the other. That alternative is the Social Teaching of the Catholic Church.

Catholic Social Teaching offers the solutions to the political, economic, and social problems that plague modern society; problems that stem from the false principles of the Reformation, Renaissance, and Revolution, and which are exacerbated by the secularization and industrialization of society that has since continued. Defending, explaining, and applying this Teaching was the business of the great Social Catholics of last century. Unfortunately, much of their work is today both unknown and unavailable.

Thus, IHS Press was founded in September of 2001 A.D. as the only publisher dedicated exclusively to the Social Teaching of the Church, helping Catholics of the third millennium pick up where those of last century left off. IHS Press is committed to recovering, and helping others to rediscover, the valuable works of the Catholic economists, historians, and social critics. To that end, IHS Press is in the business of issuing critical editions of works on society, politics, and economics by writers, thinkers, and men of action such as Hilaire Belloc, Gilbert Chesterton, Arthur Penty, Fr. Vincent McNabb, Fr. Denis Fahey, Jean Ousset, Amintore Fanfani, George O'Brien, and others, making the wisdom they contain available to the current generation.

It is the aim of IHS Press to issue these vitally important works in high-quality volumes and at reasonable prices, to enable the widest possible audience to acquire, enjoy, and benefit from them. Such an undertaking cannot be maintained without the support of generous benefactors. With that in mind, IHS Press was constituted as a not-for-profit corporation which is exempt from federal tax according to Section 501(c)(3) of the United States Internal Revenue Code. Donations to IHS Press are, therefore, tax deductible, and are especially welcome to support its continued operation, and to help it with the publication of new titles and the more widespread dissemination of those already in print.

IHS PRESS
Non est enim potestas nisi a Deo

For more information, contact us at:

mail: 222 W. 21st St., Suite F-122~Norfolk, VA 23517 USA
toll-free telephone or fax: 877-IHS-PRES (877.447.7737)
e-mail: order@ihspress.com • *internet:* www.ihspress.com

IHS Press is a tax-exempt 501(c)(3) corporation; EIN: 54-2057581.
Applicable documentation is available upon request.